ASTECHNOVA International Energy Conference 2016

Advanced Energy Innovation

E3S Web of Conferences Volume 42 (2018)

Yogyakarta, Indonesia
2-3 November 2016

Editors:

Ir. Sunarno
Liem Peng Hong
Agus Pulung Sasmito

ISBN: 978-1-5108-6592-1

TABLE OF CONTENTS

Dept. of Nuclear Engineering and Engineering Physics
Universitas Gadjah Mada

PREFACE

The worldwide energy demand is always increasing. On the other hand, the fossil-based energy resources is limited and the nuclear-based renewable energy needs much effort to optimize its utilization. Therefore, innovation in energy technology is needed. Such advanced energy innovation is not only focusing on one aspect of technology, but also it needs to consider other aspects such as environment, policies, economical point of view, etc. Synchronization and innovation from academician, professional from industry, government, and society to achieve the goal of energy independence are needed.

The ASTECHNOVA 2016 International Energy Conference is the 5th conference being held by the Department of Nuclear Engineering and Engineering Physics (DNEEP), Universitas Gadjah Mada. ASTECHNOVA 2016 filled with outstanding participants and papers, to be shared among researchers from universities, industry, government, and energy-related organizations with several topics included oil and gas, new and renewable energy, nuclear energy, energy management, green technology, environment compliance, energy policy, instrumentation and control in industry, and other relevant topic of energy. We hope that this conference can be of benefit for facilitating exchange of ideas among researchers, especially in the energy field and its related topics.

On behalf of ASTECHNOVA International Energy Conference 2016 committee, I would like to give my sincere appreciation of your supports, paper submissions, and presentations.

Thank you very much and warm regards,

Dr. Alexander Agung
General Chair

ABOUT THE CONFERENCE

ASTECHNOVA 2016
INTERNATIONAL ENERGY CONFERENCE

“Advanced Energy Innovation”

Department of Nuclear Engineering and Engineering Physics, Universitas Gadjah Mada
Phone/Fax : +62 274 580882 - Email : astechnova@ugm.ac.id
www.tf.ugm.ac.id/astechnova

Contents

Department of Nuclear Engineering and Engineering Physics, Universitas Gadjah Mada
Phone/Fax : +62 274 580882 - Email : astechnova@ugm.ac.id
www.tf.ugm.ac.id/astechnova

ABOUT THE CONFERENCE

ASTECHNOVA 2016
INTERNATIONAL ENERGY CONFERENCE
"Advanced Energy Innovation"

Short Description

Development of technologies for energy system leads to dramatic changes in energy consumption. It has never been so important to examine the stability of the worldwide energy consumption. During recent years, existing energy technologies have been showing to be inadequate due to fundamental changes in worldwide social needs and paradigm. Many basic assumptions are no longer certain and are definitely questioned. Nowadays, researchers and practitioners from industry and academia are developing advanced technology and innovation in energy system in order to solve these problems.

The ASTECHNOVA 2016 is to host communication and dissemination of proposed new approaches in dealing with technological development, as studying complexity and knowledge management. This event is the third international energy conference. In 2015 the conference was held successfully and attended by several participants and experts from Philippines, Nepal, South Africa, China, and USA.

The ASTECHNOVA 2016 -International Energy Conference- will systematically examine the following topics

- Oil and Gas
- New and Renewable Energy
- Nuclear Engineering
- Green Technologies
- Energy Management and Services
- Environmental Compliance
- Energy Policy, Education and Training
- Plant Engineering and Instrumentation & Control
- Other relevant topics on energy

Department of Nuclear Engineering and Engineering Physics, Universitas Gadjah Mada

Phone/Fax : +62 274 580882 - Email : astechnova@ugm.ac.id
www.tf.ugm.ac.id/astechnova

Rationale

Department of Nuclear Engineering and Engineering Physics, Universitas Gadjah Mada (DNEEP) is currently in the process of being a centre of excellent in energy system as part of the global institution. The most important aspects in the preparation for being a centre of excellent in energy system is the internationally scientific atmosphere in DNEEP. Department of Nuclear Engineering and Engineering Physics, Universitas Gadjah Mada should be joint to the international community in energy system with prominent researchers and practitioners from industry and academia in energy system. Therefore, it is crucial to an international conference to bring together researchers and practitioners from industry and academia for sharing their energy technological development.

In the conference, many well recognized researchers, practitioners and academia will share their experience and vision of the future technological development in energy system. Among others, there will be the scientific members, i.e. Prof. Sunil Chirayath (Texas A&M University, USA), Dr. Liem Peng Hong (NAIS Co., Inc, Japan), Prof. Agus Pulung Sasmito (McGill University, Canada), Prof. Christophe Demaziere (Chalmers University of Technology, Sweden), Hiroshige Kikura (Tokyo Institute of Technology), Prof. Sunarno, Prof. Phillip Beeley (Khalifa University, United Arab Emirates), Prof. Yonghee Kim (KAIST, South Korea), Dr. Jim Kuijper (NUCLIC, Netherlands) Dr.-Ing. Sihana (Universitas Gadjah Mada, Indonesia), Dr. Gea O.F. Parikesit (Universitas Gadjah Mada, Indonesia), Dr. Totok Soehartanto (Institut Teknologi Sepuluh Nopember, Indonesia and Dr. Jarot Setyowiyoto (Universitas Gadjah Mada, Indonesia). DNEEP believes that conducting the ASTECHNOVA 2016 - International Energy Conference - is a strategic initiative to open more networks to the prominent energy system researchers and practitioners from industry and academia in the world whose opinions and suggestions will help DNEEP to excel as a centre of excellent in energy system.

Aims

1. To bring together researchers and practitioners from industry and academia to give overviews of the state of the art of technology developments in energy system,
2. To present recent achievements of advanced technology developments in energy system,
3. To exchange ideas and experiences in the field of technology developments in energy system.

Event

The ASTECHNOVA 2016 - International Energy Conference - is proposed to take two days in November 2016 with 100 participants world wide that will present their papers, i.e.

Date : 2 and 3 November 2016

Time : 07:30 – 16:30 with four parallel scientific sessions

Venue : Eastparc Hotel Yogyakarta

Important dates for participants for this event

Extended abstract submission deadline	July 15, 2016
Notification of acceptance	July 29, 2016
Presenter registration form submission deadline	Oct 7, 2016
Full paper submission deadline (extended)	Oct 7, 2016
Conference date	Nov 2-3, 2016

Department of Nuclear Engineering and Engineering Physics, Universitas Gadjah Mada

Phone/Fax : +62 274 580882 - Email : astechnova@ugm.ac.id

www.tf.ugm.ac.id/astechnova

The detail schedule of ASTECHNOVA -International Energy Conference-, as follows:

Day 1 (2nd November 2016)

The Grand Ballroom	
Time	**Schedule**
07:30– 08:00	Registration
08:00 – 08:15	Traditional dance performance by Universitas Gadjah Mada student
08:15 – 08:20	**Welcome speech** by Committee Chairman
08:20 – 08:30	Opening by Dean of Faculty of Engineering Universitas Gadjah Mada
08:30 – 08:40	**Launching ASTECHNOVA AWARD** by Director General for Information and Public Diplomacy of the Ministry of Foreign Affairs of Indonesia and Director of Non-Aligned Movement Centre for South-South Technical Cooperation (NAMCSSTC)
08:40 – 08:50	**Photo session**
08:50 – 09:15	**Coffee break** Mini chamber orchestra by Department of Nuclear Engineering and Engineering Physics Universitas Gadjah Mada Student
09:15 – 10:00	**Keynote Speaker:** National Energy Board (DEN) - **Tumiran** "Human Resources for Energy Development in Indonesia"
10:00 – 10:15	Souvenir for keynote speaker, main speakers and representation of volunteer reviewers
10:15 – 11:30	**Challenge and Opportunities in Energy Development** **Main Speaker :** - Cubaenergia – **Curbelo Alonso** "Challenge and Opportunities in Clean Energy Development in Carribean and Latin America Regions" - Indonesia Financial Services Authority (OJK) – **Edi Setijawan** "Financial Foundation for Energy Development in Indonesia" - Dept. of Nuclear Engineering and Engineering Physics, Universitas Gadjah Mada - **Rachmawan Budiarto** "Education, Research and Community Development Activities on New and Renewable Energy within Universitas Gadjah Mada" **Moderator :** Haryono Budi Santosa
11:30 – 12:00	Discussion
12:00 – 13:00	**Poster Session (3rd Floor)** **Lunch** at Verandah Restaurant (1st Floor)

Department of Nuclear Engineering and Engineering Physics, Universitas Gadjah Mada

Phone/Fax : +62 274 580882 - Email : astechnova@ugm.ac.id
www.tf.ugm.ac.id/astechnova

Scientific Parallel Session

Time	Dahlia Room	Tulip Room 2
	Renewable Energy	**Oil and Gas and Other**
	Chair : **Hiroshige Kikura (Tokyo Institute of Technology)**	**Chair :** **Agus Arif (UGM)**
13:00-13:25	#13Multilevel Diffuser Augmented for Horizontal Axis Wind Turbine (Fariz Qashidi Putra)	#33Anti Surge Control of Centrifugal Compressor at PT. Pertamina Ep Asset 2 Field Pendopo (Muhammad Lukmanul Hakim , Balza Achmad, et.al)
13:25-13:50	#18Purification of Biodiesel Using Activated Carbon Produced From Cocoa Pod Husk (Dan Sulaiman)	#15Approach to Topographic Map, hydrogeological and Environmental Isotopes Analysis to Delineation of Watersheds in Drought-prone Areas (Aris Sunantyo)
13:50-14:15	#22The Development of Heterogeneous Catalyst CaO/NaOH From Waste of Green Mussel Shell for Biodiesel Production(Hadiyanto, Budiyono, Istadi, et.al.)	#43Microscopy of Alloy Formation on Arc Plasma Sintered Ods Steel (Bandriyana, Arbi Dimyati, et.al.)
14:15-14:40	#11Renewable Energy: Numerical study of the effect of geometry variation on the performance of innovative design wind speed enhancer (Theodorus Tio Wibowo)	#29An Optical Current Sensor Based on an SMS Fiber Structure (Agus Muhamad Hatta, Ika Novitasari, et.al.)
14:40-15:00	**Poster Session (3rd Floor)** **Coffee Break**	

Time	Plant Engineering and Instrumentation & Control	Plant Engineering and Instrumentation & Control
	Chair : **Balza Achmad (UGM)**	**Chair :** **Totok Soehartanto (ITS)**
15:00-15:25	#4Numerical Study of A Thermosyphon Cooling System: Film Condensation (Filian Arbiyani)	#6Post-Installation : Insight of Coastal-Area Society in Hybrid Power Generator of Pantai Baru, Yogyakarta (Widhyharto)
15:25-15:50	#28Performance Evaluation of Three-Bed Adsorption Refrigeration Cycle (Aep Saepul Uyun, Kamarudding Abdullah, et.al)	#39Energy Monitoring System for Existing Buildings in Indonesia (Sentagi S Utami , Faridah, et.al.)
15:50-16:15		#17The Study of System Planning of Assets Security on Marine Current Renewable Energy (Juniarko Prananda)
17:00-19:00	**Welcome Dinner** Invited by NAMCSSTC **(The Grand Ballroom)**	

Department of Nuclear Engineering and Engineering Physics, Universitas Gadjah Mada

Phone/Fax : +62 274 580882 - Email : astechnova@ugm.ac.id
www.tf.ugm.ac.id/astechnova

Day 2 (3rd November 2016)

Time	Dahlia Room	Tulip Room	Magnolia Room
		Renewable Energy	Plant Engineering and Instrumentation & Control
		Chair : Kusnanto (ASEAN Centre of Energy)	Chair : Awang N. I. Wardhana (UGM)
08:00-08:30	**Morning meeting for Day 2 chairs and speakers (Iris Room)**		
08:30-08:55		#25Development of Biobriquette From Nutmeg Seed Shells (Rukmana Idris, Ahmad Tawiequrrohman, et.al.)	#30Design of Hydro Power by Using Turbines Kaplan on The Discharge Channel Paiton 1 And 2 (Totok Soehartanto, Alvin Kurniawan Sosilo, et.al.)
08:55-09:20		#40Reactive Extraction of Reutalis Trisperma Seeds(Nunung Prabaningrum)	#34Potential Analysis of Ex-Coal Mining Land as Energy Storage Type Pumped Hydroelectric Storage in Kutai Kartanegara, East Kalimantan (Yollanda Zilviana Devi, Rachmawan Budiarto, et.al.)
09:20-09:45		#10Urban Photovoltaics and Online Energy Monitoring System: Development in UGM's Kinanti Dormitory (Rachmawan Budiarto)	#35Development of Fuel Debris Inspection Methods Using Air Coupled Ultrasound (Ari Hamdani, Hiroshige Kikura, et.al)
09:45-10:05		#27Influence of Temperature Variation to Optimize Biogas Purification on Water Scrubber System (Arief Abdurrakhman , Roekmono, et.al)	#47Techno-Economic Evaluation of Carbon Capture Storage Ready for Coal-Based Power Generation In Indonesia(Usman)
10:05-10.30	**Poster Session (3rd Floor) Coffee Break**		

Department of Nuclear Engineering and Engineering Physics, Universitas Gadjah Mada
Phone/Fax : +62 274 580882 - Email : astechnova@ugm.ac.id
www.tf.ugm.ac.id/astechnova

Day 2 (3rd November 2016)

Time	Renewable Energy	Energy Policy, Education and Training	Nuclear Engineering
	Chair : Nunung Prabaningrum(UGM)	Chair : Singgih Hawibowo (UGM)	Chair : Sihana (UGM)
10:30-10:55	#49Energy in Lao PDR (Keomany VHANSALY)	#9Influential Factors to The Industrial Consumers' Smart Grid Adoption (Zheng Ma)	#2Study on A Hybrid Nuclear System: Light Water Reactor with Fossil Fueled Superheater (A.F. Wibisono)
10:55-11:20	#50Energy Development in Vietnam (Tran Thi Thu Trang)	#26Nuclear Security Training: A New Approach for Educating Nuclear Technology in Indonesia (Ferdiansjah, Faridah, et.al.)	#3Fuel Loading Pattern Optimization with Constraint on Fuel Assembly Inventories Using Quantum-Inspired Evolutionary Algorithm(M. Rizki Oktavian, Alexander Agung, et.al.)
11:20-11:45	#51Bioenergy in Cambodia (YANG Sokchea)	#48Measuring Indonesia's Energy Security (Qodri Febrilian Erahman)	#5Mass and Energy Balance Analysis of Methanol Production Using Atmospheric CO_2Capture With Energy Source from PCMSR (Andang Widiharto)
11:45-12:00	**Closing ceremony (Iris Room)**		
12:00	**Lunch** at Verandah Restaurant (1st Floor)		

International Scientific Committee

Prof. Sunil Chirayath (Texas A&M University, USA)

Dr. Liem Peng Hong (NAIS Co., Inc, Japan)

Prof. Agus Pulung Sasmito (McGill University, Canada)

Prof. Christophe Demaziere (Chalmers University of Technology, Sweden)

Prof. Hiroshige Kikura (Tokyo Institute of Technology, Japan)

Prof. Sunarno (Universitas Gadjah Mada, Indonesia)

Prof. Phillip Beeley (Khalifa University, United Arab Emirates)

Prof. Yonghee Kim (KAIST, South Korea)

Dr. Jim Kuijper (NUCLIC, Netherlands)

Dr.-Ing. Sihana (Universitas Gadjah Mada, Indonesia)

Dr. Gea O.F. Parikesit (Universitas Gadjah Mada, Indonesia)

Dr. Totok Soehartanto (Institut Teknologi Sepuluh Nopember, Indonesia)

Dr. Jarot Setyowiyoto (Universitas Gadjah Mada, Indonesia)

Local Organizing Committee (Universitas Gadjah Mada, Indonesia)

General Chair	: Alexander Agung
Publicity & PR	: Sentagi Sesotya Utami
Financial Chair	: Faridah
Treasurer	: Amrih Lestari, Safrudin
Publication Chair	: Agus Arif, Balza Achmad
TPC Chair	: Awang Wardhana, Anung Muharini, Singgih Hawibowo
Program Chair	: Susetyo Hario Putero
Logistic Chair	: Widya Rosita
Sponsorship & PR	: Haryono Budi Santosa, Rachmawan Budiarto

ASTECHNOVA 2016
International Energy Conference

Dept. of Nuclear Engineering and Engineering Physics
Faculty of Engineering
Universitas Gadjah Mada

+62 274 580882
astechnova@ugm.ac.id
www.tf.ugm.ac.id/asctechnova

edp sciences

Statement of Peer review

In submitting conference proceedings to E3S Web of Conferences, I certify to the Publisher that I adhere to the **Policy on Publishing Integrity** of the journal in order to safeguard good scientific practice in publishing.

1. All articles have been subjected to peer review administered by the proceedings editors.
2. Reviews have been conducted by expert referees, who have been requested to provide unbiased and constructive comments aimed, whenever possible, at improving the work.
3. Proceedings editors have taken all reasonable steps to ensure the quality of the materials they publish and their decision to accept or reject a paper for publication has been based only on the merits of the work and the relevance to the journal.

Title, date and place of the conference

Astechnova 2016
International Energy Conference
2 -3 November 2016
Eastparc Hotel, Yogyakarta, Indonesia

Proceedings editor(s):

Prof. Ir. Sunarno, M.Eng., Ph.D (Universitas Gadjah Mada, Indonesia)
Prof. Agus Pulung Sasmito (McGill University, Canada)
Dr. Liem Peng Hong (NAIS Co., Inc. Japan)

Date and editor's signature

17th June 2017

E3S Web of Conferences **42**, 01001 (2018) https://doi.org/10.1051/e3sconf/20184201001
ASTECHNOVA 2016

Multilevel Diffuser Augmented for Horizontal Axis Wind Turbine

Fariz Qashidi Putra[1], *Dani* Rifai[1], *Kutut* Suryopratomo[1,*] ,and *Rachmawan* Budiarto[1]

[1] Departement of Nuclear Engineering and Engineering Physics, Universitas Gadjah Mada, Indonesia

Abstract. Indonesia is an area with the low and fluctuating wind speed. Therefore, the implementation of the wind turbine to generate electricity become ineffective and economically unprofitable. Diffuser Augmented Wind Turbine (DAWT) is the augmentation technology in wind turbine which could increase wind speed flow that arrives on the turbine blade. The working principle of the diffuser is to create a difference in pressure inside and outside the diffuser. The pressure inside the diffuser is lower than the pressure outside so that the wind will be accelerated into the diffuser and wind speed will dramatically increase at the inlet of the diffuser. This study will be presenting a modified design of diffuser augmented wind turbine (DAWT) by designing multilevel diffuser with additional inlet curvature and flange. This research aims to evaluate the amplification of flow velocity profile around the diffuser that has been engineered. The numerical study is performed using computational fluid dynamic (CFD) to obtain the highest ratio of speed increment. The verification of numerical initial condition is validated by comparing the result of validation with experimental data available in the literature. The result shows that a ratio of increase in speed is 2.08 times higher than conventional wind turbine over 4 m/s inlet velocity. In the equation of a wind power output, the wind speed is proportional to the cubic power of its wind power output. Therefore, the utilization of dual-stage diffuser device in wind turbine would give significant increment on the power output of wind turbine.

1 Introduction

Indonesia, as a developing nation, still depends on fossil energy to fulfill almost the whole need of national energy. In 2025, Indonesian Government has several regulations in terms of reducing the use of fossil energy and supporting renewable energy development such as wind energy, solar energy, geothermal energy, ocean energy and bioenergy. Geographically, Indonesia is surrounded by two continents and has the second longest the coastal line in the world. As a result, lots of area with a high intensity of wind located in Indonesia definitely. Moreover, wind energy has certain utilization to generate electricity by using wind turbine. Unfortunately, recent research and study which is conducted by Lembaga Penerbangan dan Antariksa Nasional (LAPAN) proved that the development of wind energy technologies in Indonesia facing some vital obstacles that block the development and people's interest in exploiting wind energy as an electricity generator, such as low-speed wind velocity distribution in Indonesia at around 2.5 – 6 m/s which is under cut-in speed specification for most wind turbines and huge fluctuation wind velocity in Indonesia not only in speed degree but also in quantity degree. One of technology that can be applied is with the additional shrouded wind turbine or Diffuser-Augmented Wind Turbine (DAWT).

The diffuser is an additional device of wind turbine that would increase wind speed when the wind comes into the wind turbine. Principal work of diffuser is to make a pressure difference between its inlet and outlet. Inlet pressure lower than outlet pressure, so that the wind speed being accelerated into the diffuser and the wind speed would increase enormously in an inlet. The use of flanged diffuser produces low-pressure area in an outlet because of the existence of vortex formation and draw more a lot of mass flow into diffuser [1], [2]. Furthermore, the utilization of diffuser with inlet curve could ease the wind to flow into the diffuser and will create streamline from wind flow [3]. The research about wind turbine was started from 1920 by Alexander Betz. This research was reviewed by Sanuki (1950), Iwasaki (1950), Lilley and Rainbird (1956), Kogan and Seginer (1963), Gilbert et al. (1978), Gilbert and Foreman (1983), and Igra (1981). The next research still uses diffuser structure which is cover the wind turbine with additional features in the form flanged on the outlet sides. The research about diffuser design was being conducted by Matsushima et al. (2006), Abe and Ohya (2004), and Ohya & Karasudani (2010).

Several ways have already conducted to upgrade the power produced by a wind turbine in low wind velocity area, such as by undertaking modification to the wind turbine on its rotor and diffuser (Kosasih, 2012). Another possible approximation is manipulating the wind velocity so that local velocity in the inlet of a wind turbine would be higher [4].

Ohya et al. were comparing three different types of hollow structures that might be installed in wind turbine. Those are a nozzle, cylindrical and diffuser (Fig. 1). It was found that an enormous increase in wind speed was

* Corresponding author: kutut@ugm.ac.id

obtained by diffuser type. Afterward, the diffuser had been engineered by adding several parameters such as the angle of inlet shroud opening, hub ratio, the height of flanged, inlet curvature, and center body length. The height of flanged in the diffuser was being variated with diffuser's diameter (h/D). The experiment result showed that the power of wind turbine with diffuser had enhanced up to 4-5 times more than the conventional one. The specifications of diffuser prototype (class of 500 W) are L = 1.25D, D = 0.72 metres, h = 0.5D and the diameter of rotor is 0.7 metres.

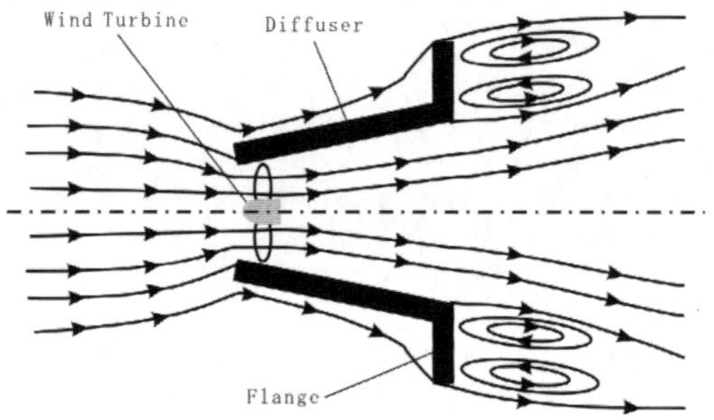

Fig. 1. Schematic view of flow mechanism around a flanged diffuser (Abe &Ohya, 2004).

Optimization of Diffuser Augmented Wind Turbine (DAWT) was conducted by Kannan et al. [5]. The study was comparing four design of diffusers by using ANSYS Fluent. The best diffuser generates local wind velocity of 61.25% increment that of the approaching wind speed. It has 160 of diffuser opening angle with the addition of splitter inside the diffuser. The length and the opening angle of the splitter are 0.5 meters and 40 respectively.

Tito et al. [6] were discussing the effect a variation of flange angle on diffuser DAWT performances using CFD. The existence of flanged caused the vortex formation at back side of the flange. Quantitative data showed flow performance in an interior diffuser with top increased velocity augmentation of the flanged diffuser is angle 90° by 9.82 m/s or 1.96 times freestream wind velocity with L/D ratio for about 7.7.

In the equation of a wind power output, the wind speed is proportional to the cube (n3) of a wind power output. If we could increase the wind speed by exploiting the fluid dynamic nature around a structure or topography, the power output of a wind turbine can be increased substantially.

Thus, in the present work, carries out a study using Computational Fluid Dynamics (CFD) to assess the effect of the velocity speed-up ratio inside the diffuser. This simulation emphasizes the geometric design of stage diffuser development which has the highest of ratio increase in speed.

2 Methodology

The study was performed using CFD numerical simulation methods. By definition, CFD is a focused study that might predict a fluid flow, heat transfer, chemical reactions, and other phenomena by completing mathematical equations numerically. One advantage of using CFD is able to conduct experimental and analytical approach [7]. Numerical simulations were performed on a multilevel diffuser assumed without using a rotor. It aims to obtain a position with a leverage ratio of maximum speed $((U_{max}/U_0)$ that could be used as a reference position for laying the rotor inside the diffuser.

2.1 Numerical Condition

This study was using a numerical method that is used to correspond to the finite volume method (FVM), using ANSYS 17.0 Student Version software. Reynold-Averaged Navier-Stokes (RANS) equation is used as a solution. In general, RANS equation for the steady state and incompressible turbulent flow, so that the equation for conservation mass and momentum are:

$$\frac{\partial U_i}{\partial x_i} = 0 \quad (1)$$

$$U_j \frac{\partial U_i}{\partial x_j} = -\frac{1}{\rho}\frac{\partial P}{\partial x_i} + \frac{\partial}{\partial x_j}\left\{v\left(\frac{\partial U_i}{\partial x_j} + \frac{\partial U_i}{\partial x_i}\right) - \overline{u'_i u'_j}\right\} + F_i \quad (2)$$

with ρ, P, U_i, u_i and v are represented density, average static pressure, average velocity, turbulent fluctuation, and kinematic viscosity respectively. F_i is the expression body-force which is represented a load [1]. In this research, the simulation was done without considering the rotor, so that $F_i = 0$. It aims to get a position at a maximum ratio increase in speed as a reference for placing the rotor. The quantity $-\overline{u'_i u'_j}$ is known as Reynolds Stress Tensor or Reynolds-averaged value.

2.2 Geometry and Computational Condition

The simulation is used a fluid material such as air at the temperature of 27°C (Table 1). The study sets the flow as axisymmetric steady flow, the segregated solver is proposed, and velocity formulation is absolute whereas the turbulence model used based on the result from validation process. The boundary conditions that are used in the validation phase numerical model could be seen in Table 2. The solution algorithm uses SIMPLE (semi-implicit method for pressure linked-equation) with Second Order Upwind scheme for turbulent parameter and Second Order for the pressure with 10^{-5} of the convergence criterion. The solver that being used is pressure-based and steady.

Table 1. Air properties at 27°C

Property	Value
Density (ρ)	1,1765 kg/m^3
Dynamic Viscosity (μ)	1,8538 $\times 10^{-5}$ kg/m-s

Table 2. Properties boundary condition of simulation

Boundary Condition	Dimension (m)	Type of Boundary Condition
Left	2.0	Velocity-inlet
Right	2.0	Pressure-outlet
Bottom	3.0	Axis
Top	3.0	Wall (no-slip)

To validate the model, we were used the geometry of diffuser proposed by Abe and Ohya (2004) (see Fig. 2). The diameter at the inlet diffuser (D) is 0.20 m, diffuser opening angle (ϕ) is 4° and the free stream velocity (U_0) is 5 m/s. As we could see in Fig. 2 and Fig. 3, the CFD simulation performed with a 2D axisymmetric assumption so that the diffuser design is being designed along the center line only. Axisymmetric has the ability to describe the modeling of a 3D object in rotational symmetry.

The computational condition for diffuser model research exists on Fig. 4 and Fig. 5. The symbol D, h, θ, dan L represent diameter size of inlet diffuser, the height of flange, angle of the diffuser, and diffuser length respectively, while subscript 1 and 2 are represented first stage diffuser and second stage diffuser respectively. A model is built with a diameter of inlet diffuser (D) and flange height (h) are 20 cm and 0.5D respectively. The opening angle of the first diffuser, the opening angle of second diffuser, the length of the first and second diffuser are 40, 80, 0.25D and 1.25D respectively.

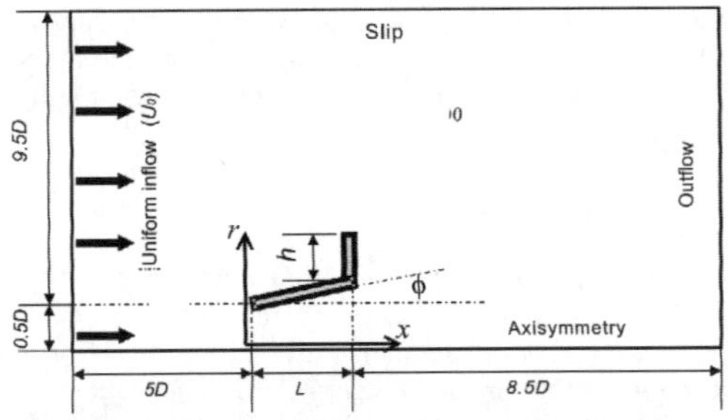

Fig. 2. Geometry and computational conditions (Abe and Ohya, 2004) for validation.

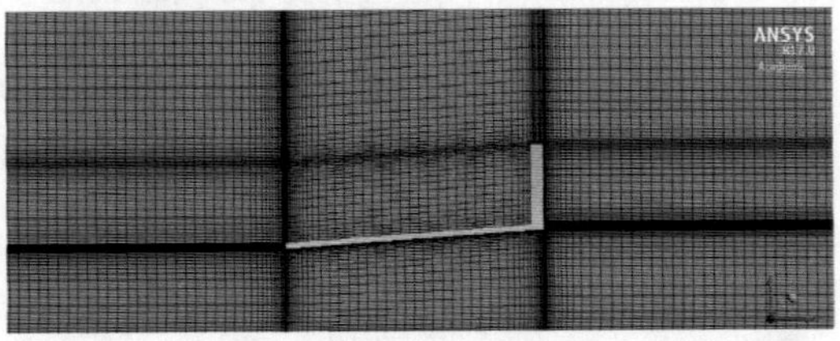

Fig. 3. The result of meshing process of diffuser model by Abe & Ohya (2004) for validation.

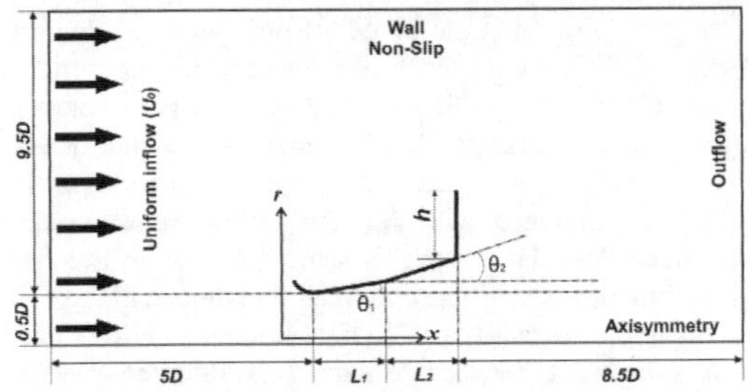

Fig. 4. Geometry and computational conditions of multilevel diffuser

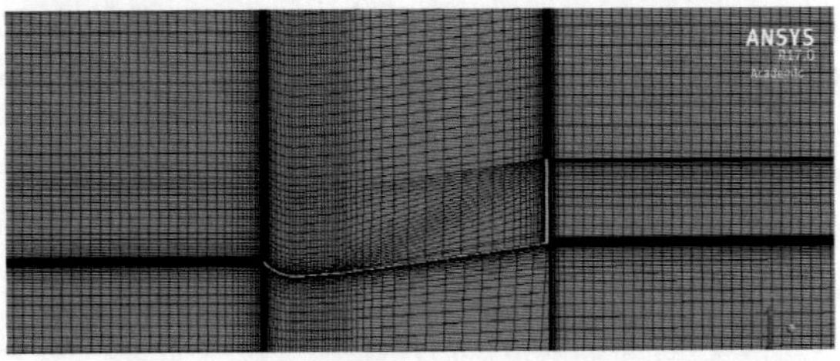

Fig. 5. The result of meshing process of the multilevel diffuser.

3 Result And Discussion

3.1 Validation of Numerical Model

Validation is used to obtain a numerical model that will be used as a reference in the simulation model of research in order to get the accurate results. Validation is done by using secondary data. The results of the validation are tried to approach the experimental results of research diffuser by Abe & Ohya [2] as close as possible. Validation results are shown in Fig. 6. X/D is the ratio of the diameter inlet diffuser and position, U/U_o is the speed ratio distribution along the axis and U_{max}/U_o is the ratio of the highest speed on axis (U_{max}) with velocity freestream (U_0), as seen in Table 3.

Table 3. Comparison of U_{max}/U_0 values of the results of validation and secondary data [1]

Turbulence Models	The results of Validation (U_{max}/U_0)	Abe & Ohya Experiment [1] (U_{max}/U_0)	Relative Error
Spallart-Allmaras	1.61	1.60	0.36%
ske SWF	1.51	1.60	5.83%
ske EWT	1.53	1.60	4.24%
rngke SWF	1.49	1.60	6.79%
rngke SWF	1.48	1.60	7.24%
rke SWF	1.45	1.60	9.24%
rke SWF	1.47	1.60	8.42%
skw	1.41	1.60	11.69%
sstkw	1.48	1.60	7.38%

Fig. 6 shows that the model of turbulence of Spalart-Almaras able to approach the results of experiments conducted by Abe & Ohya [1]. Types of solver as used in this study are pressure-based, axisymmetric and steady. Turbulence models that being tested are Spalart-Allmaras, K-epsilon standard with Standart Wall Function and Enhanced Wall Treatment, K-epsilon RNG with Standart Wall Function and Enhanced Wall Treatment, K-epsilon Realizable with Standart Wall Function and Enhanced Wall Treatment, K-omega standard and K-omega SST.

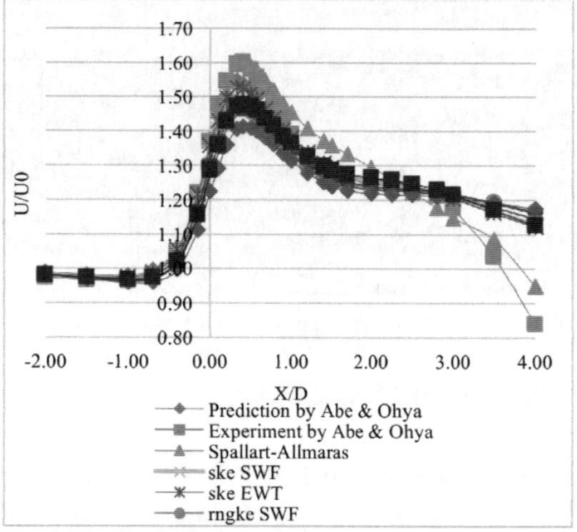

Fig. 6. The distribution of velocity ratio (U/U_0) along the axis upon the validation of numerical models for various turbulence models.

Fig. 6 illustrates the wind speed distribution along the axis on the various turbulent model. Inlet position of the diffuser in $X/D = 0$ and the outlet position in $X/D = 1.5$. The wind speed would increase when the wind come into the diffuser. The existence of low pressure inside diffuser leads to accelerating the wind speed. The geometrical shape of diffuser also affects this large increase.

3.2 Accuracy of Validation Result

This research uses numerical simulation with CFD rather than experimental. Therefore, we need to figure out the accuracy of the simulation results obtained by looking at a few parameters, such as mesh independence study, mesh quality, and mass imbalance.

3.2.1 Study of Independence Mesh

Mesh independence study was conducted to determine the number of mesh that will be used in validation and simulation research. The change of mesh independent in cells number is not a significant effect on the simulation results.

From the graph in Fig. 7, the value of the output of the speed (U_{max}) starts to level out when the number of the mesh are approximately 60000 cells. Therefore, the number of mesh that can be used to validate the simulation models are roughly 60000 cells.

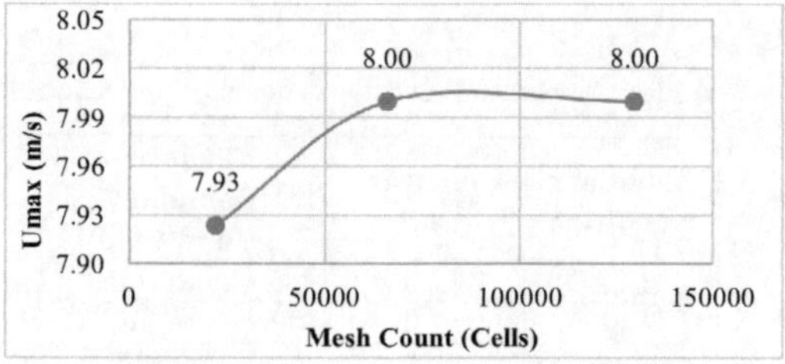

Fig. 7. The study of mesh independence for validation numerical model.

3.2.2 Mesh Quality

Mesh quality could be determined during the process of meshing by the parameters of skewness and orthogonal quality. Mesh with poor quality will lead to inaccurate simulation results and/or simulation requires a longer time to achieve convergence, even it may cause the simulation diverges. The value of skewness and orthogonal quality of research diffuser are 0.003 and 0.999 respectively. Thus, the mesh quality of simulation results for the validation implied in the excellent category.

3.2.3 Convergence Criterion

Convergence criteria determine the numerical convergence that would occur in the numerical solution. When the convergence is achieved, the numerical solution is unable to change again, due to the overall equation (mass, energy, momentum) on each of the cells has had fulfilled the convergence tolerance. The value of convergence criteria in the default state is 10^{-3}. In this study, the value of the convergence criteria was changed to 10^{-5} for speed and continuity. The smaller the order is used, the error level in the settlement process will be smaller.

3.2.4 Mass Imbalance

Another important parameter to determine convergent simulation results is to examine the overall imbalance mass flow (mass imbalance), which is obtained from the difference between incoming mass flow and mass flow out. The appropriate value of mass imbalance should be less than 1% to meet the criteria of mass imbalance. The value of mass imbalance on the simulation results in the validation could be seen in Fig. 8. The magnitude of mass imbalance is 2.00 x 10^{-7} or <1%.

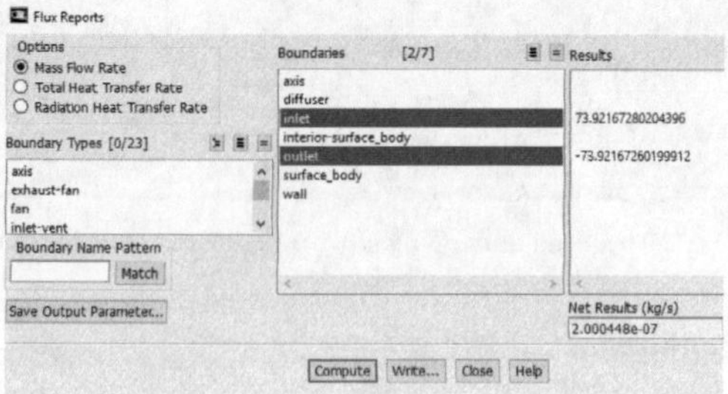

Fig. 8. The result of a mass imbalance in the validation of numerical models.

3.2.5 Multilevel Diffuser Simulation

In this paper, the geometry design was used a multilevel diffuser with flange and an addition of inlet curvature. This study aims to present a simple numerical simulation using CFD to find the ratio of increase in speed from each geometrical model.

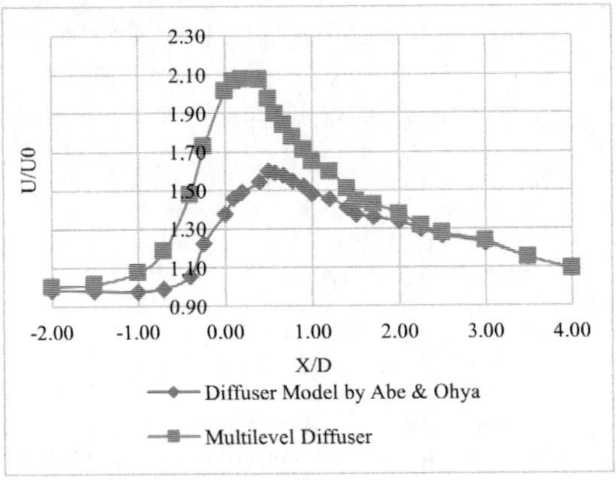

Fig. 9. Comparison of speed increment ratio distribution along on-axis for the multilevel diffuser model and diffuser model by Abe & Ohya

Fig. 9 shows the comparison of speed increment ratio distribution along on-axis for the multilevel diffuser model and diffuser model by Abe & Ohya. The ratio of the increase in speed generated by multilevel diffuser model is very significant, relative to diffuser model by Abe & Ohya. The ratio of the increase in the maximum speed (U_{max}/U_o) reaches 2.08 times higher at 4 m/s velocity inlet.

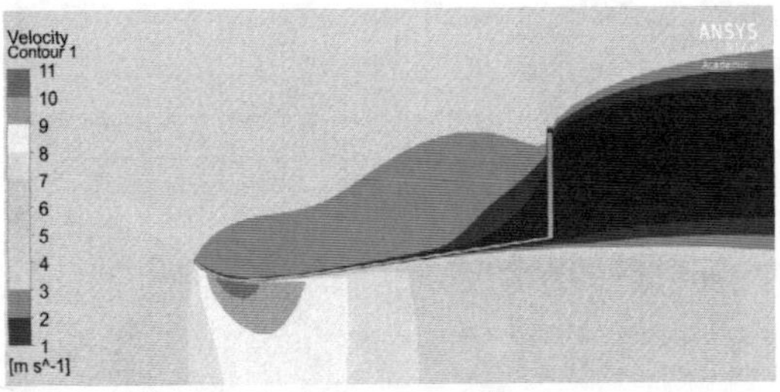

Fig. 10. Contour plot of the velocity of multilevel diffuser model.

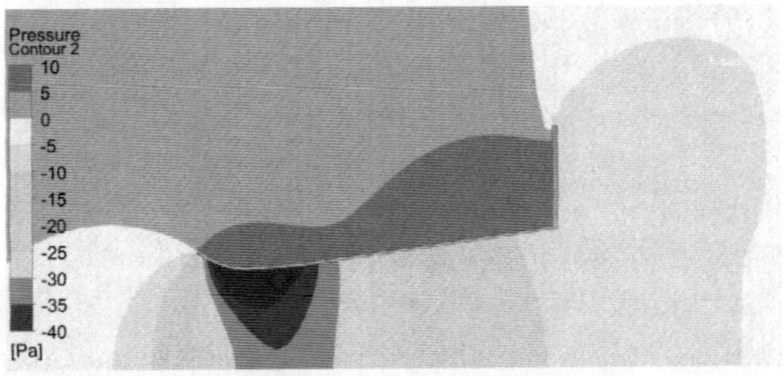

Fig. 11. Contour plot of the pressure of multilevel diffuser model.

Fig. 10 and Fig. 11 show the contour plot of velocity and contour plot of pressure from multilevel diffuser model are represented with its color degradation. Red indicates a high value, while blue indicates low value. In the diffuser, velocity increases while the pressure decreases. The increase in the highest speed is obtained at the lowest pressure. As seen in Fig. 10, there is a reddish orange color on the inside near the diffuser inlet showing the highest speed. While in Fig. 11, there is a wide dark blue on the inside near the diffuser inlet in which shows the lowest pressure.

Table 4. The distribution ratio of U/U_0 at some measurement points along the axis.

X/D	U/U_0	
	Multilevel Diffuser	Diffuser Model by Abe & Ohya
-2.00	1.00	0.98
-1.50	1.01	0.98
-1.00	1.07	0.97
-0.70	1.19	0.99
-0.40	1.48	1.05
-0.24	1.73	1.22
0.00	2.01	1.38
0.10	2.07	1.46
0.16	2.08	1.49
0.20	2.07	1.54
0.40	1.97	1.60
0.50	1.89	1.59
0.58	1.84	1.58
0.68	1.78	1.55
0.78	1.71	1.52
0.90	1.65	1.48
1.00	1.60	1.46
1.20	1.51	1.41
1.40	1.44	1.37
1.50	1.43	1.36
1.70	1.38	1.33
2.00	1.31	1.30
2.25	1.27	1.26
2.50	1.23	1.22
3.00	1.15	1.15
3.50	1.09	1.09
4.00	1.05	0.95

Table 4 illustrates the velocity distribution of the X-axis at some point in the ratio of X/D. D is the diameter of the diffuser inlet. The highest speed ratio is in a position $X/D = 0.16$. This position could be used as a reference position to locate the rotor on further research experiments.

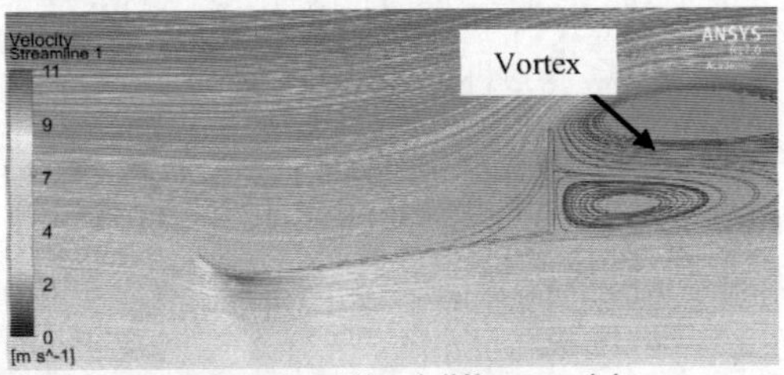

Fig. 12. Streamline of multilevel diffuser model.

https://doi.org/10.1051/e3sconf/20184201001

3.3 Accuracy of Multilevel diffuser simulation

3.3.1 Study of Independence Mesh

The study of independence mesh was conducted to determine the number of mesh that will be used in validation and simulation research. The change of mesh independent in cells number is not a significant effect on the simulation results. Therefore, the number of mesh that can be used to validate the simulation models are around 60000 cells.

3.3.2 Mesh Quality

The value of skewness and orthogonal quality are 0.016 and 0.999 respectively. Thus, the mesh quality of simulation results for the validation counted into the excellent category.

3.3.3 Convergence Criterion

In this research, the value of the convergence criteria was changed to 10^{-5} for speed and continuity.

3.3.4 Mass Imbalance

The value of mass imbalance on the simulation results for multilevel diffuser model could be seen in Fig. 13. The magnitude of the imbalance is 2.73 x 10^{-7} or <1%.

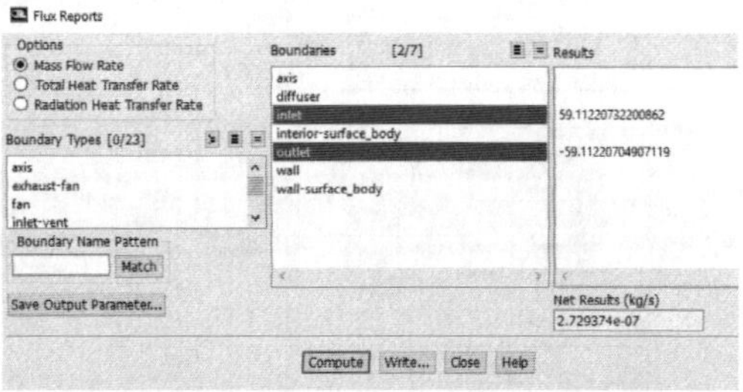

Fig. 13. Mass imbalance results in validation of numerical models

4 Conclusion

Numerical study of the design geometry of multilevel diffuser with flange and an addition of inlet curvature applied for a wind turbine has been performed using ANSYS 17.0 Student Version. In this study, a model turbulent of Spallart-Allmaras was used to solve the equations RANS. It is according to the results of the validation process that is able to approach the experimental results Abe & Ohya (2004). The ratio of increase in the maximum speed (U_{max}/U_o) that could be reached by the multilevel diffuser model is 2.08 times higher at 4 m/s velocity inlet and the recommendation position to put the rotor of wind turbine is in $X/D = 0.16$.

The results of testing the accuracy of the simulation results refer to the study of mesh independence, mesh quality and mass imbalance shows that the CFD simulation results in research of multilevel diffuser wind turbines are accurate and acceptable. However, this research mainly relying on numerical simulation and so does need a further experiment to be tested and verified in the real condition.

References

[1] K.I. Abe, Y. Ohya, J. Wind Eng. Ind. Aerod **92**, 315–330 (2004)

[2] Y. Ohya, T. Karasudani, A. Sakurai, K.I. Abe, M. Inoue, J. Wind Eng. Ind. Aerod **96**, 524–539 (2008)

[3] Y. Ohya, T. Karasudani, Energies **3**, 634–649, (2010)

[4] D. Wahyu, Purwanto,. Geometrical Variation Analysis of Diffuser Performance on Diffuser Augmented Wind Turbine. ITS. Surabaya.

[5] T.S. Kannan, S.A. Mutasher, Y.H.K. Lau, (2013). Design and Flow Velocity Simulation of Diffuser Augmented Wind Turbine Using CFD, **8**(4), 372–384. (2011)

[6] A. Tito, H. Wibowo, R.A. Wahyuono,. Numerical Study of Geometrical Performance and Diffuser Design for Improving the Performance of Diffuser Augmented Wind Turbine, 14(2), 90–96. doi:10.9744/jtm.14.2.90-96. (2013)

[7] T. Jiyuan, G.H. Yeoh, C. Liu, *Computational Fluid Dynamics: A Practical Approach* (Elsevier, Oxford, 2008)

E3S Web of Conferences **42**, 01002 (2018) https://doi.org/10.1051/e3sconf/20184201002
ASTECHNOVA 2016

Post-Installation: Insight of coastal-area society in Hybrid Electric Power Systems (HEPS) of Pantai Baru, Yogyakarta

Derajad S. Widhyharto

Departement of Sociology, Faculty of Social and Politic Sciences, Universitas Gadjah Mada, Indonesia

Abstract. The Hybrid Electric Power Systems project is a pilot renewable energy project of the Indonesian Ministry of Research and Technology since 2010. Hybrid power (windmill and solar cell) was chosen because of their huge heat and wind potential in coastal area of Pantai Baru. Post-installation is critical, considering the existence of alternative energy was expected to bring changes in the life of coastal people nearby. Participatory Action Research (PAR) was the method used to do deeper exploration of the changes happened after the hybrid electric power installation in Pantai Baru. Coastal society had a bigger dependency on natural resources and secluded from any technology development, including the hybrid power itself. The use of PAR method reversed the top down logic by the Indonesian Ministry of Research and Technology with the bottom up logic, adjust the people's needs which processed from the grassroots. This research was conducted in three years, using three approaches; exploration, reflection, and action. Through these approaches, it was possible for the researchers to understand the actual condition of coastal society and chose the women groups as the main informants as well as the main group which got the biggest benefits of the Hybrid Electric Power Systems (HEPS), they were capable of giving fast responses on the changes happened through the culinary, tourism, fishpond, etc. This condition emphasized both the central role of women and the knot of all dynamics of the HEPS installation. In conclusion, women groups positioned themselves as the transformation agent of the usage of this renewable energy around HEPS Pantai Baru. As the main beneficiaries, women groups had improved their capability to adapt. It increased the income, consumption level, etc. Despite the success of the adaptation, they were still failing their position as the agents and users of the renewable energy. Their existence was limited merely to objects/beneficiaries, instead of being the subjects/actors of the HEPS technology. This study is essential as a data baseline, prepare the policy of technology intervention as well renewable energy campaign.

1 Introduction

Nowadays, energy is included as one of the main issues discussed in various international agreement. This happened as a form of energy crises anxiety. It is not excessive though when we see the correlation of energy and human being. Human is highly dependent on energy. The process of industrial revolution changed the way people live. Mechanization of all human activities became dominant. Therefore, energy as the main power of human's living tools is absolutely necessary.

So far, the world's energy supplies are dominated by the non-renewable ones. This type of energy come from fossil which produced oil, coals, and natural gas. It needs a long time for a fossil to transform into oil, that is why it is called as non-renewable energy.

The anxiety of the energy crises has emerged throughout the world as there are many estimations said that fossil energy would soon completely decreased. Its high consumption, all for households, industries, military, and others will bring us to the energy crises. The urgency to create renewable energy became clear throughout the world. It was proven when renewable energy became the major issues on several international conferences. In the early June 2004, delegations of 145 states attended the international conference called Renewables 2004 in Bonn, which produced 'Political Declaration' and 'Action Program'. They were previously invited by Gerhard Schroeder in 2002 when German Chancellor delivered a speech at the UN World Summit on Sustainable Development in Johannesburg. The fact that one of the three biggest industry countries took a step in this initiative remarked the renewable energy issues as a part of the world politics.

Optimism in renewable energy emerged in many countries. Those optimism and energy security seemed to be too early. If we take a look at the non-renewable energy (fossil fuel), its consumption rate is relatively increasing. Non-renewable energy consumption is high above the renewable one. In 1990, based on the International Energy Agency, global consumption of fossil fuel[1] reached 530,000,000 metric[2] 'ton oil equivalent' (TOE). It reached 8,130,000,000 ton, there was 44 percent growth in 12 years only. Still, in the same period, renewable energy market segments increased from 1.04 to 1.38 billion metric feet, or 33 percent increased. The consumption gap between fossil fuel and renewable energy growth 4.59 – 6.74 billion tons in 12 years.

This continuous growth of energy consumption is a paradox of states' commitments on energy issues. They committed to energy providence, but on the other side, there is the continuous growth of energy consumption each year. Fossil fuel is still considered as the most

[1] (crude oil, coal, natural gas)

[2] (Metric 'feet' is the international unit of conventional measures for all types of energy);

efficient resources. Differs from renewable energy produced from atom or solar which need advanced technology and high cost. Meanwhile, fossil fuel has a limitation, and it is inevitable. The explanation above matched to Herman Scheer (2007) argument on "Energy Autonomy". Scheer tried to explore various intersections in renewable energy issues. This is important because each day the non-renewable energy start to be diminished. Whilst, renewable energy is still becoming a discourse. This process will bring us into energy disaster if there is no committed action.

Environmental issues previously were the major issues in several conferences and in several countries. Developed countries had planned to develop renewable energy to minimize the environmental damages in the future, such as global warming and greenhouse effect. This concern then encouraged several countries to develop renewable energy usage which eco-friendly and apply eco-friendly lifestyle. Some countries have specific regulations or policies about the renewable energy usage. It emerges the crucial investment, besides the economic investment.

However, renewable energy issues bring out political economy motives. Some parties use the renewable energy concept as an economic investment opportunity which is profit oriented. The coalition between market and government shifted the renewable energy focus. Their actions are not based on environmental and social issues anymore, whereas they focus on making profit for certain parties related to market and government. Renewable energy policies are not that effective. There are always certain purposes during the implementations. The news as well as remarks about it remain as discourses. Market and governments interests play a part here; they are afraid of losing profit when they shift the usage of fossil fuel to renewable ones.

Society development depends on the way it fulfills the needs of energy. Electrical power is the major energy needed by people's daily lives. Therefore, it is basic needs for a production process which affect the economic growth of society. For example, the electricity entry point in a certain area will not solely give them facility in modern agriculture, but also facilitate the modernization of other production processes. The government also give credit to the farmers, by co-operating with some banks which have the branch offices in rural areas.

Indonesia has problems on distributing energy for its people. There is unfair energy distribution in some areas in Indonesia. As an archipelago country, our National Electricity Company or Perusahaan Listrik Negara (PLN) as the main distributor faces challenges in the distribution process. It is making sense to find an alternative way to answer these problems, especially in some remote areas. The Indonesian government is then using renewable energy as one of the solutions.

Through the central government, local government gets the authority to develop renewable energy which expected to be able to fulfill the people's electricity needs. One of the government regulations about renewable energy is Constitution No. 30 Year 2007 about energy. In the verses number 20 and 21 of that constitution written that "(central) government and local government makes effort in provision and management of energy from new sources and renewable energy sources which is addressed corporations, permanent enterprises, and individuals are able to get facilitation and/or incentive from (central) government and local government based on its given authority for certain amount of time until they reached the economic values."

Indonesian Ministry of Research and Technology as the related stakeholder of energy issues had done several types of research about renewable energy in some areas in Indonesia. The coastal area is one of research objects in the early of 2010 because it is one of the hardest areas that PLN need to distribute electricity. One of the areas which used as experiment and research is Pantai Pandansimo Baru. HEPS technology was tried to be installed there because of the similar topography (soil characters). It is also confirmed by the huge wind potential which capable of moving the HEPS turbine.

Moreover, coastal people livelihood and characters are very diverse, such as farmers, merchants, cattlemen, fishermen. The researchers took fishermen as a sample as they constantly lived nearby the coast. Fishermen's lives in Pantai Baru coast, Dusun Ngentak, were not that different from other fishermen's. Even the stereotype about fishermen's habit of wasting money after sailing also happens to them. Fishermen tended to spend their money only in one day to buy things they want at once. They only cared of what they need that day, barely think about the next days and so on. Most of the fishermen in Dusun Ngentak were the lower middle class and still using traditional lifestyle.

Most of them were still having a problem in conserving their seafood. They have to buy ice at the relatively high price and even take long distances to get it. Thus, their profit did not optimally distribute as they take the additional cost for production (fish preservation). Fishermen's lives get worsened by the growth of groceries and fuel prices needed for their job. Uncertain and low-income got them harder to fulfill their daily needs. Thus, it was too far for them to reach the welfare.

Moreover, in the middle of unpredictable weather and climate as the effect of global warming made them somewhat afraid to go sailing. Bad weather such as huge tide was one of the reasons they choose to stop sailing sometimes. Therefore, they did not get anything for a certain amount of time. Without income, they find it hard to buy fuel. Sometimes, the cost of fuel was higher than their income from sailing. As a survival, particularly in the middle of the unpredictable weather like those days, they had their own strategies to overcome it, such as being farmers, or merchants in the beach area and other things. Coastal people had prepared themselves to become seafood culinary entrepreneurs. However, their efforts sometimes were interfered by lacking electricity, so that their small restaurants only open until the afternoon, they could not make it by night. They were also facing the lack of clean water. Hence, the distribution of electric power was a major necessity for Pantai Baru people as the infrastructure to support their development.

The program to conform their needs of energy made to empower them, and it was an innovative idea. As it is known, maritime and fishery development had not shown much significant impact yet. It was proven on how fishermen were still on the edge of poverty. Therefore, it was a good action when government and related Ministry wanted to create alternative solutions and thoughts which meant to empower local people. Thus, they made fundamental changes in the fishermen's perspective, that they might not merely depend on the climate. Fishery management needs to be emphasized on fishermen's self-management. Considering Indonesia as an archipelago country, Indonesia needs more capable people to accommodate all those natural resources as well as responsibly maintain its sustainability. Besides, with that proponent natural potential, people might be able to meet their own needs independently.

After the HEPS technology trial which had been running for five years in the coast of Pantai Baru, people start to see the changes in their lives, both on economic and social aspects. Even, HEPS which is located on the coast of Pantai Pandansimo, Dusun Ngentak, is dubbed as the biggest HEPS in Indonesia. Thus, this research was conducted to know the impact of HEPS installation towards the local people and how far the installation affects the dynamics in Pantai Baru society.

2 Research Method

This research was conducted for three years (2012-2014), with the assumption that the duration is effective to the concept of PAR. HEPS was installed on trial in 2010-2011 so that the two years after that assumed as the post-installation momentum. Two years could show the adaptation process of society. PAR is orientated to three approaches, exploration, reflective, and action McIntyre, 2008). The first year was the explorative period. Researchers did exploration on the potential and social challenges which are predicted to happen on the field. It began with the actor mapping and even made database about people's lifestyles nearby the HEPS. This was important to map the informants and identify values, norms, social status, and social roles emerged in the field. Considering the trip from campus or researchers' residents to Pantai Baru took about an hour in the morning and 1.5 hours in the evening. Thus, researchers chose to not stay in the location, instead visit the location for an effective week, once a month regularly. In the first year, researchers made some approaches to all of the related stakeholders such as representatives from Ministry of Research and Technology, HEPS managers, workshop, and operators. The result of this first year was actors map, strategic issues, and social lives narrative of local people.

The second year was the reflective year. The approach was changed into reflective to get direct interactive ways between researchers and local people. This action was important to both strengthen and help findings triangulation from the previous year. In this second year, researchers decided to do on-the-spot FGD and in-depth interview to the chosen informants. It was essential as the research was not meant to interfere people's activities, as it was believed there would not any authentic data/answers when we asked them to allocate specific times for the discussion or interview. Hence, we followed their daily activities for a week once a month. There was an assumption that a week was the best period to see the pattern, referred to the sequence days on one week. The result of this second year was transcripts and narratives from many perspectives, problems, and application map of HEPS. In this second year, researchers decided to make coastal women near HEPS Pantai Baru as the main informants.

From the second year, the development of HEPS had already mapped and inflicted environmental effects around it. Several affected parties in Dusun Ngentak were the fishermen group and merchants group. Both of the group were affected directly and mostly because of their locations were nearest to the HEPS construction area.

Another party affected by HEPS construction was people who work as farmers and cattlemen. HEPS construction did not bring impact as big as it is to them brought to people who domiciled and did economy activities in the coastal area. The least affected group of HEPS installation was the people who were not included as merchants, fishermen, farmers, and cattlemen. People worked as civil workers, employees, or factory labors and others did not get any direct impact of the HEPS.

The third year, it had reached the action phase. Research process had already found the map of issues, problems, and HEPS utility relations. Thus, persuading people to do actions towards both the issues and their lives as coastal people is the next goal this year. In this third year, it was also prepared a workshop for women group of Pantai Baru coast to encourage them to think about renewable energy in their surroundings. This workshop was held in their houses or shops. Even, women group were also asked to do a comparative study tour to another place which was relevant for the development of their entrepreneurship and the utilization of renewable energy as the focus. The result was a map of the entrepreneurship development opportunities as the utilization of the renewable energy of HEPS Pantai Baru.

Those activities in the third year showed that HEPS had an impact on the society, including the women of Dusun Ngentak who are the main focus of this research. Women of Dusun Ngentak experienced all changes, from the HEPS pre-installation whereas they were still focused on domestic sector then after its installation in Pantai Baru they moved forward to the public sector.

Generally, not all data in the research used in this article. The writer tried to use several data thought to be relevant to the issues about the post-installation impacts.

According to researchers, there were some weaknesses of using PAR as a method, such as 1) it was not easy to build trust with informants in a relatively short time, considering the condition where researchers did not stay in the location every day, 2) from the people's point of view, this method could represent grassroots voices, but this method could not reach the higher level to change or affect the policy-making process at the same time, 3) This method needed sufficient time to do data identification and extraction, as researchers relied on the limited numbers of field notes, transcripts, interview records. Hence, researchers and team consolidated the data after we finished the interviews to rewrite on the log book and reckoned on our memories from the last interviews we did on the field with the informants.

3 Literature Reviews

Feenberg (2002), in the preface of his book "Technology and Transformation : From Marxian to Radical Critique" emerged the argument of Harry Braverman that the Capital concept of Marx is not a single concept. According to Braverman, there are two major critics of capitalism which related to one another. Feenberg specifically divided it into two conceptions, namely property theory and labor process theory. While property theory related to the economy analytical process, labor process theory is associated with the sociological process as the main base.

Labor process theory defined the way of capitalist organized the workers or labor as deskilling or abolishment of their authority in creating something. Its relevance to technology, the usage of the machine as replacement of men power enables them to decline the numbers of workers. Thus, in this context, besides their loss of authority, workers also lose their creativity as they played the role of machine operators. The usage of the machine or deskilling will indeed cut the cost of wages they should pay significantly. In the other side, this will also influence the political process in the society where deskilling used as capitalist's hegemony basis, both in the work environment and society in general.

The people's ignorance and inability in understanding this production process in the system, making them the easy object of capitalist hegemony. All this time, workers thought that the knowledge they had in the specific area is their subjective rationality as a human. In fact, production mechanism changes their subjective knowledge as an object of authority, wherein the end subordination is not something to be enforced anymore, instead, it is known as a consequence of technology usage.

"A new"law"of economic life has arisen with the new technology, a law that commands"fitness of the laborer for varied work, consequently the greatest possible development of his varied aptitudes."However, by its very nature capitalism requires just the opposite, an ignorant and docile labor force tied to highly specialized tasks. This is the" absolute contradiction between the technical necessities of Modern Industry, and the social character inherent in its capitalistic form" (Marx, 1906:1, 533-534 in Feenberg, 2002).

The statement above tell us that the use of technology upon the men power had reversed the logical thoughts of the industry. If previously, there is a belief that right persons for the right jobs, capitalists have reversed it with people who are the foolest and obedient as the most favored person with skills needed for the jobs. It is exactly differed and contradicted to the technical necessity of modern industry as well as the social characters relevant to any forms of capitalism.

Talk about the relevance of technology, Marx could be said unable to explain the relations between the work process and technology. Marx still thought that to reach socialism, the production force needs to be separated from the capitalist production relations. This perception never talks about the position of technology in his way to separate production force from the mechanism of the production process. George Lucas in Feenberg (2002) said that Marx's perception was not technology determinism. He assumed technology and work relations as production mode.

Besides, Marx was seen as ambiguous in explaining the position of technology in his conception of capitalism critics. This ambiguity can be influenced by his effort to reject romanticism about technology instrumental values which would overtake him. Hence, Marx finally limits his critic about technology merely on the abuse of machine. He also wrote that people who reject those rational reasons as fools because they did not fight against capitalist but merely fight against the machine as themselves. In a scrutiny which not only focuses on negative influences, critics would be spread to the whole aspects of machine and technology. Thus, people who fight against or criticize technology without knowing the limit would be ignorant, as a person, he will not be able to detach himself from any form of technology.

Variance or separations of "workers and technology" as itself, in fact not simply discharge Marx from critics. Since there is no "technology as itself " because technology comes up or only exist when there is reasonable context to support its existence. In another word, there will be no machine or technology if there is no job to finish. That is why technology relevant to the function of "use" . Even Feenberg argues that the usage of word "machine" become ambiguous. Since, the usage of this machine word can be directed to any form of technology used by people, whether in the context of capitalization or not.

Based on Marx, usage of technology is considered as bad when it determined into these indicators : 1) the actual purposes they want to achieve by using technology. 2) how is the technology used apart from its purposes, 3) technical principles in the production process will influence in forming technology, as well as the duration of work hour, work atmosphere, and other problems related to work safety.

In this article, technology is not assumed as a neutral entity and accessible for everyone. It is a thing needed to be learned through education and rational planning. While the access to education is not available for everyone, thus, the capitalist's attraction to technology involving its design, innovation, choices, and usage method. Although it is never be written, Marx critics to capitalize actually offend technology either. Whereas, based on the design of this critic, capitalist technology actually was formed and influenced by other capital aspects, such as management as the controller.

In the end, critics to technology rise as a part of the technology development which cannot be separated from interests, where social power upon people's necessity is one of the reasons why technology exist. Since then, this article perceives technology as the dependent entity, instead of a dependent variable to the social system, as a purpose which is formed by the dominant class, as well as a subject from the creating effort of new hegemony in society.

3.1 Technological Rationality

Marcuse's technological rationality theory in Feenberg (2002) was influenced formally as formal rationality is less dependent on values. In his theory, Weber argued that rationalization will clearly prosper the capitalists. It is inevitable that the control over technology, whether it is in the past or now is the major source of power. Marcuse argued in his technological rationality theory that technology does not care about people. Technological rationality is the tools to control people. The control is not solely to the individuals and the equipment but either to the individual's freedom itself. Marcuse argued more that where there are production activities, there must be domination towards the individuals. Domination might occur through the managerial system or work division system so that it will be more efficient. Domination system occurs so that workers will never take control of the company. This concept is similar to the alienation concept by Marx, but its domination is not revealed through regulations, however, unconsciously is internalized by workers.

3.2 Power/Knowledge

This article also uses Foucault's statement which quoted by Feenberg (2002) as a correction to Marcuse's theory which is somewhat bias. Foucault also believed that power and domination have openly done. Social control which had applied by Foucault is more about control of time efficiency to achieve discipline. Foucault believed that technology is not value-free, there is partiality towards capitalists. Truth and power are not separable, where there is power, there will be the truth. For example, leaders are always right. Foucault said research about human would not be certainly influenced by the situation. Human's cognitive and social power are similar to Marcuse's duality about enforcing rationality which using vehicle assembly as an example. Foucault argued that control/knowledge is power to dominate individuals as both subjects or objects.

The subject here is defined as individuals as themselves, whilst object is defined as individuals with others, such as technology or else. The system used here is not enforcing, but in the other way can make individuals obey rules automatically. Technology is one of the tools to rule individuals. When there is trouble to individuals, it will be assumed to be technical trouble and not enforcement. Later this thing is structured, and when there is trouble, the capitalists see how the workers fix it.

3.3 Technological Configuration

While the green revolution was taking place (1960-1970), stakeholders in this case government expected new varieties in agricultural products, particularly staple food. The expectation emerged because the fundamental perception of Green Revolution is increasing agricultural products through some following steps : provide the irrigation system, optimize chemical fertilizer, usage of pesticide, and usage of quality seeds. Although, mechanization of agricultural area is included in the agricultural modernization process. It is not the only one; there is still any other technology such as pesticide, fertilizer, etc.

The green revolution and agricultural modernization, farmers, are demanded to speed up their production process. So that, after the harvest season, farmers need to get the area fertilized and ready for planting. One of the essential instruments in the agricultural process is a tractor. Farmers who have tractors will successfully improve their lives, with following indicators : enhancement of clothing, meals, houses, and recreations. This encourages other farmers who do not have a tractor yet to buy ones, let alone the market offers various kind of agricultural equipment.

Furthermore, technology existence is potentially rising socio-economical stratification. People who used to have social integration become have dysfunction, for example, farm labor who do not own any field loss their job because technology replaces their position, encourage them to move to the urban area and occupy workforce in informal sectors.

Based on the explanations above, some of the research findings by M.J. Campbell (1989) shown that impact of technology in several countries, particularly in their rural areas create significant changes, below.

3.3.1 Changes in family roles

The family who had experienced mechanization in the process of harvesting process spend less time in the field because the process is faster compared to a family who has not go through mechanization process.

3.3.2 Impact on children growth

Before the modernization and technology, children in rural areas are expected to contribute in domestic sectors in their early ages : get water, deliver food, etc. Nowadays, the family tends to realize the importance of education for children, both boys and girls.

3.3.3 Impact on labors

Some perceptions about the impact of agricultural mechanization in rural areas are increasing numbers of employee and worsening of poverty. Yet, data indicates that there is some misunderstanding in the perception about labors. It happens because of several following factors.

a. Although modernization and mechanization had already happened, the necessity of labors in agriculture is not declining. Farm labors are still needed for planting and weeding process, as its harvest and planting period are much faster.
b. There are transportation facilitation and infrastructure which enable unemployees to find jobs in other places.
c. The government of rural areas provide new vacancies, at least in short period of time, and provides vacancies for untrained or semi-trained workers.

3.3.4 Changes in works and skills

Agricultural mechanization encourages villagers to understand the equipment they need to use. At least they know the maintenance procedures and simple reparation. This knowledge will lead the farmer family has skills to operate the tractor in a safe way. Furthermore, this process will be used as learning process by one of the family members who have specialization in one certain thing, such as tractor reparations, irrigation tools replacement.

3.3.5 Relations between family and society

Agricultural mechanization and the life improvement increase public services as well as people's expectations. The residential standard has enhanced, demands on water and electricity have become primary needs of each family. The entrance of private sector in a rural area has changed rural people's perspective; then they started to become more independent to fulfill their own facilities that government cannot provide. Families who have capital has increased their influence in rural society. Then, their capital becomes social modal to decide the policies in the rural area.

3.3.6 Relations to the other market sector

The entrance of private sectors in rural areas formed new markets. Some dominant areas in this new market forming process are:

Savings and loans institutions, both government or privately owned ones.

a. Sales and services institutions of agricultural machines.
b. Political and religious institutions.
c. Entertainment provider institutions for rural people, social services, and health clinics.

3.3.7 Mobility of Families

The mobility of families, whether it is traditional or the ones who already mechanized expand their perspective about the world they have never seen before. As people meet the world, they have experience interacting with new societies. It positively impacts the way they see their rural area and lead them to be more open to changes. While the negative side is they tend to break their cultural norms they used to obey all this time.

3.3.8 Changes in aspirations and expectations

The family which is experiencing mechanization has widened necessity compared to the one who is still traditional. There is an anxiety in modernized rural people that they might not be able to meet their needs. It is not based on intuitive reasoning. Instead, it comes from their past experiences. People's needs that used to be simply made for survival has enhanced into – mostly – fulfillment of life improvement facilities, such as further education access up to university. People's expectation of life can be divided into four categories : 1) high standard of life : electricity and water, 2) proper houses : houses which build properly, 3) better infrastructures and 4) recreations for people.

3.3.9 Changes in family economy

The improvement of the economic base among rural people is the impact of their great expectation towards the facilities they want. Even, to fulfill their needs, there is a tendency to take the loan as a way to increase their capital or improve their economic base.

In the end, critics to technology rise as a part of technology development itself which cannot be separated from interests. Social power upon people's needs become one of the determinants the creating of technology. This article considers technology, not as independent elements, instead one of the variables dependent on the social system, a purpose which formed by dominant class, as well as subject from the creation process of the new hegemony in the society.

Evolution occurs in the technology cannot be seen as an autonom process, yet have to be associated with interests and social groups.

This approach saw the labor division and technical conception are unable to compare to the development of workers' capacity individually. Because they who are part of the worker class never put interest in technology because their discussions are still questioning welfare. It proves the argument of Feenberg (2002) at the beginning of this literature review that there is an incompatibility between what workers want and the reality had done by capitalists in the field. The production process is designed to be a control system by the dominant class rather than fulfillment mechanism of the worker's rights. In this Pantai Baru HEPS case, coastal women played a part as the workers of technology.

4 Research Findings

In response to the Pantai Baru HEPS construction, here are some actors who earn benefits from the HEPS construction, one of them is a group of women. Women are quite vulnerable to changes around them, particularly their domestic situation at homes as well as the environment. When there is new element come in (read: HEPS), it will give major influences to women condition, such as her job and income. After the installation of HEPS, women which are previously identified to be solely in domestic sectors, finally come to the public sector (Amirrudin, 2012).

Women's issue in the economy is not far from poverty (Anggraeni, 2010). Therefore, giving them the opportunity to participate in economic activities and other more chances are the right way, whereas women can work outside of their homes, contributed to their families' income.

Women's role in the informal economy sector is one of the significant roles in upgrading their status in the social structure of Dusun Ngentak. Here is following socio-economical changes emerged post-installation of Pantai Baru HEPS.

4.1 Fishery Group

This group is found since there was a program of fishery ponds construction nearby the location of the Hybrid Electric Power Systems (HEPS) initiated by the Ministry of Maritime and Fishery. The formation of this fishery group was initially made as women empowerment activities in Dusun Ngentak. Fishery ponds were then being managed by women of Dusun Ngentak which members of the fishery group called Kincir Mina Mulya Group.

Another purpose of its establishment was making use of empty land around the windmills, and making it as new side projects for the women of Dusun Ngentak to get additional income. In the management process, the Ministry of Maritime and Fishery did a workshop to improve people's knowledge about the benefits and fishery management system. That workshop was held for women (mothers) of Dusun Ngentak which belong to Kincir Mina Mulya Group. The workshop was held like direct socialization about "how to be lucky fish farmer". Lucky here was meant to be a fish farmer who can manage the fingerlings well so that they would get the best result as well. In the process, Ministry of Maritime and Fishery gave them initial capital aid, namely fingerlings and fish feed. Furthermore, they also held training session (practical training) and comparative studies to get more potentials and knowledge about fishery management in the coast of Pantai Baru.

HEPS installation in the coast of Pantai Baru had been the main livelihood for the existence of fishery group Kincir Mina Mulya. The fish ponds relied on the electric power produced by the HEPS. The energy was used to move the water pumps so that the water will flow to the fish ponds. Meanwhile, the ownership status of the ponds based on the ownership of the land. Thus, in the managerial operations of the fishery pond groups, they responsibly manage all the things such as maintenance and profits together.

In the process, when the harvest time came, not suddenly all the farmers of Kincir Mina Mulya successfully got profits. Some of them were failed. It happened when the farmers could not maintain their ponds well. There were found some of the ponds unmaintained so that they were unable to produce fishes continuously. There were some abandoned ponds and were not optimally used. Otherwise, the farmers who could make use of the fish ponds as their economic potentials. They could take benefits from the Ministry's aid so that they got profits of it.

The good intention from the Ministry of Maritime and Fishery, to actualize the women empowerment program in Dusun Ngentak, through the establishment of fishery ponds in Pantai Baru coast also got both pro and contra reactions from people.

"Yes, just like the realization of these ponds constructions, when all the tools and materials to construct them were totally afforded by the government the people's sense of belonging to those ponds must be less than what we expected. Automatically, their responsibility is weak either, because they didn't contribute any capital. So that, we see many failures in the first year. It because the fishery farmers' low sense of belonging to maintain and do the fish breeding." (citizen of Dusun Ngentak)

According to pro groups and people who have positive responses to this program, saw these ponds construction by Ministry of Maritime and Fishery as benefits. Since those ponds could be used as their second job which took not much of their time but gave them additional income for their daily needs. However, there were dissents, particularly people who gave a negative reaction , one of them was Jaman[3] (50 years old), one of the fish merchants in the tourism spot of Pandansimo Baru. He said his disagreement to the installation of HEPS.

According to Jaman, the HEPS installation process in the coast of Pantai Baru was not initialized with proper socialization to people. So that, in the process, not all of the people could accept it. For example, construction of fishponds, that program was dubbed as a waste of money. That program seemed to be unsustainable.Whereas, it was planned to be not merely as additional income, but also support the lives of less productive fishermen when they were not able to sail. Yet, to Jaman, those fishponds did not successfully solve the lack of fish stocks.

Besides, another thing which tribulates fishery farmer's struggle was the feed supplies. Fingerlings and fish feeder from the Ministry of Maritime and Fishery only came for once at the beginning of the program. Thus, farmers need to feed the fishes by themselves to keep the fishponds sustainable, maximized the initial capital aid from the ministry. The price of factory-made feeder for fish was really high so that it became a burden to them. Meanwhile, high production cost discouraged many of them to leave this kind of activities as they thought it was not prospective anymore. In the end, the ponds left abandoned.

[3] Interview 11.03.2012, 10.30 a.m.

Therefore, it is necessary to find the alternative fish feeder which is affordable and have the same nutrition as the one which made by the factory. One of them is maybe reproduced worms in 1x1 meters in a medium containing cow's waste and worms so that it is hoped their problem on fish feeder issues will be solved.

4.2 Agricultural Group

Geographically, although it is located on the coast of Pantai Baru, Dusun Ngentak areas were dominated by fertile agricultural land. Thus, most of the people there work as farmers. It frequently used as their main activity for a living. However, the high production cost and unpredictable climate change often lead them to failure. The failures made the farmers suffer a big loss. According to them, not having any profit from the harvest meant they would not have any meals to eat. Thus, it was necessary to find an alternative way for them to pay their loss after the crop failures, at least to afford their daily needs. While waiting for the harvest time which could take a long time, they could do other things to earn money. Anything, as long as they could fulfill their needs which were not slightly simple.

At the beginning of the HEPS installation in Dusun Ngentak, farmers were one of the groups who reacted negatively toward the plan. HEPS construction certainly needs time and much land in the process. So that, it was necessary to give wide spare land for the HEPS. This condition was seen as a threat to farmers. A Certain amount of agricultural land was converted for the sake of the HEPS installation. However, most of the land used by farmers was part of the Sultan Ground, which meant it all belonged to the Sultan of Yogyakarta, and they were afraid and upset when they asked them to give it back to the Sultan. It happened because the land conversion from agricultural land into HEPS area (although not all of the land) was not only involved them as the farmer but also related to their families' lives.

Nevertheless, as time went by, through some effective and prevalent approaches as well as socialization, people finally accepted the HEPS installation in a positive way. As Mujiyo, the Leader of Farmers Group[4] said people then realized and accepted the decision to install HEPS in Pantai Baru coast. It is not merely because the land was part of Sultan Ground, they then believed that HEPS would make a change and give benefits in their lives continuously.

The hopes were finally answered. People of Dusun Ngentak, particularly the farmers, got to benefit from the HEPS installation; they could cut the production cost. It happened because of the irrigation system. Before the HEPS existed, farmers needed 60,000 IDR to irrigate their fields. It was already covered the fuel cost, solar and diesel which they needed to rent for. While after the installation they only needed 10,000 IDR. HEPS had a crucial role in the irrigation system in Dusun Ngentak farming area. Farmers could use the electric power to operate the pumps to water their lands.

Despite the domination of men in agricultural works, the women of Dusun Ngentak also actively participated. It could be seen as the women regularly helped their husbands in the field, from the seed planting until the harvest time. Although the men still dominated the process, women's active participation in agricultural activities helped farmers to press the production cost and accelerate the process of seed planting, so that they did not need to wait for harvest time any longer.

4.3 Culinary Group

After the installation of HEPS in Pantai Baru, trading became the new favorite livelihood of Dusun Ngentak people, especially the women. They began to sell seafood; seafood culinary entrepreneurship was the new most favorite job for the women there. The opportunity was taken by women who previously sell snacks in front of their homes and housewives. The emergence of these culinary entrepreneurs began after the HEPS. HEPS became a magnet which attracts people to visit Pantai Baru. The tourist enthusiasm encouraged people of Dusun Ngentak to finally process the seafood they got from their fisherman husbands. Culinary entrepreneurship was expected to add the family's income. So far, there were at least 80 (eighty) families listed as a member of culinary entrepreneur groups in Pantai Baru coast.

The culinary entrepreneurship phenomenon in the coast of Pantai Baru which was dominated by women proved that women could exist and contribute in public sector. For example, Suharyati (25), one of the culinary entrepreneurs, said that before she joined the culinary entrepreneur's group, she was only active in the domestic sector and only helped to look after their cattle. Since she became a culinary entrepreneur, Suharyati found a way to express her skills and interest in culinary, as well as help her family's economic condition.[5]

Although the location which was used for the culinary outlets was not their private lands, the merchants did not get any charge or paid rent cost for it. But, they had to pay 1,000 IDR per week to the leader of the culinary entrepreneur's group, Jamali. It was used as the cost for their own developments, such as training for seafood processing innovation. The culinary outlet construction was a part of the economic development program of HEPS installation. Based on its size, the outlets divided into several types. The first type (I) is the biggest one and has the biggest size, namely 12 x 25 m. The second type (II) is 6 x 9 m, and the third type (III) is 5 x 6 m. The people who could occupy the outlets and do economy activities there was only the people of Dusun Ngentak, the outsiders were not allowed. It was because the ones who could take direct benefits from the HEPS were the nearby people first. It was also expected that women empowerment could be implemented as it planned before.

The benefits of HEPS post-installation for the women of Dusun Ngentak, especially who worked in culinary were described below.

[4] Interview, 11.03.2012, 12.30 p.m.

[5] Interview 11.03.2012, 12.30 p.m.

4.3.1. Crystal ice

Crystal ice production became prominent for the culinary entrepreneurs post-HEPS-installation. It was easier than for them to get consumable ice. The distance between the ice producer, HEPS, and the outlets is quiet adjacent and made it easier for the entrepreneurs. On the other hand, the price was really affordable and boosted the sustainability of this business. It made the entrepreneurs enable to cut their production cost to the price of the ice so that they got more profits. One big box of ice is 10,000 IDR. If they bought it as retail, it cost 1,000 IDR. Meanwhile, they used to buy ice far from HEPS for 2,000 IDR.

4.3.2. Electricity

The outlets got electricity from the HEPS. Although it was relatively small (the electric current), it had helped many outlets there. On the weekend and on holidays like Saturday and Sunday, the activities are going on until the evening, depending on the visitors. They were not only used for restaurants, but the outlets were also used for social meetings and activities, such as prayers, sports activities, or many other things until the evening. The HEPS takes an essential role in the business.

Nonetheless, the entrepreneurs of Pantai Baru were still expecting for the development of HEPS. It was because they hoped that HEPS could provide greater energy, not only to street lightings but also for other necessities. As Anis (41 years old), one of the entrepreneurs of Pantai Baru, she hoped that HEPS would be able to provide electricity for her cooking equipment such as rice cooker, magic jar, blender, and so on so that she could serve her customers faster.

4.3.3. Tourism Group

Post-installation of HEPS in the coast of Pantai Baru emerged several new work groups in Dusun Ngentak society. As it explained before, there were some groups emerged after the installation of HEPS such as fishery, agriculture, and culinary groups. As an addition, they also had the tourism group who emerged as the great tourists' enthusiasm to visit Pantai Baru post-HEPS-installation. HEPS was tourism attraction so that it made Pantai Baru became more well-known. HEPS made Pantai Baru as a beach with an educational message, named as a technology-based tourism spot. The HEPS windmills of Pantai Baru were developed as Techno Park and became an edutourism icon on the coast of Pantai Baru (Widhyharto, 2015).

Tourism group was born as the HEPS developed involving clothing merchants, public toilet keepers, ATV renters, kids pool renters, and parkingmen. Previously, tourism group of Pandansimo Baru already existed, but not many people interested in it. It was shown on the small number of visitors in Pantai Pandansimo Baru. After the HEPS windmills installed, the tourists were growing each day.

Yanti (29 years old) as one of the clothing merchants in Pantai Baru said that before the HEPS installed, Pantai Baru was deserted so that there were no economic activities around the beach. Yanti was only a housewife before; then she could help her family's economy by being clothing merchants in Pantai Pandansimo Baru area. Previously, she was culinary entrepreneurs, but to her selling clothes was more prospective compared to culinary. According to her, HEPS took important roles in opening business opportunity as well as its sustainability. Before HEPS exist, there was no activity in the evening, even in the afternoon, the beach looked so quiet. Thus, after HEPS was installed, Pantai Baru became alive and could help people for a living.[6]

Lastri (29 years old) also felt the benefits from the HEPS installation in Pandansimo Baru coast. Lastri worked as public toilet keeper after the HEPS installed, said that she felt the positive change in the economic growth of Dusun Ngentak people. In a half day, Lastri could earn 70,000 IDR. Tourism group of Pantai Baru which also involving the ATV and kids pool renting, was completed by many tourism facilities for the visitors. It helped them to improve the beach's tourism attraction because the location which was near to HEPS.[7]

4.3.4. Women on Post Installation

Post HEPS installation in the coast of Pantai Baru, had made a great social change in the women economic condition of Dusun Ngentak, Bantul, Yogyakarta. In conformable to the first plan of the HEPS installation, which was to empower women. Women of Dusun Ngentak had been given the opportunity to contribute in economical aspect. The chances were proven on the emergence of business groups which take a chance after the HEPS installation, such as fishery, agriculture, culinary, and tourism groups. In each sector, women had an active role in developing those sectors.

Culinary, for example, most of this sector was overtaken by women. Women had space to express their skills and interests in processing foods. Women were the main target of this culinary development program, considering their cooking skills, the culinary sector was dubbed as their new arena to express women existences in public sectors. Easy access to get the main as well as the complementary ingredients to be processed as delicious seafood were one of the good reasons to develop this business. The seafood culinary business was supported by its strategic location which was on the coast. Women could directly help their husbands to sell their seafood through served them into delicious dishes.

So did the women who worked in the fishery sector. Using the opportunity to be fishery farmers, women who diligently work on it could get profits and continuously breed the fishes. Although, sometimes they experienced failures, and suffered great loss. So that, it came the Indonesian proverb, "besar pasak daripada tiang" or in

[6] Interview, 11.03.2012, 12.30 p.m.
[7] Interview, 11.04.2012, 15.30 p.m.

English "bigger pole than the peg", meant there was unbalanced budgeting lead to the greater production cost than the profit, at least they could take lessons from each experience. Step by step they would be able to measure the market potential to get profit. Women of Dusun Ngentak had the courage to try and learn. Yet, not all of them get the great profit, at least through this fishery business they could decline their boredom of household routine. It was also helped them to fulfill their daily needs. At least, through this two sectors, we could see the existence of HEPS in Dusun Ngentak had to impact in making changes of women's social and economic condition. The emergence of new business groups after the installation was the prove. Although it only provided a small amount of power.

Before the installation, most of the women of Dusun Ngentak were housewives which only active in domestic sectors only. Being wives and mothers is the usual daily role of those women. Husbands were the only one who worked for a living and fulfills the family needs. That explained why they lived in a barely sufficient condition. Although, not all of them, most of the women of Dusun Ngentak felt less productive if they were only active in the domestic sector and only helped their husbands sometimes if it is necessary.

Nonetheless, since they became more independent, ran their own businesses, women helped husbands to do economic activities for their families. When they were working, women could earn additional income, so that she could manage their household income to make some saves for the future. Actually, those four factors could be done by one woman to be followed by other women. It was because there was easy access, provided allowance/subsidy for the initial capital, strategic location, followed by business opportunities. But, the limited space made not all of the women get the opportunities (Widhyharto, 2014).

HEPS installation and the tourism development in Pantai Baru not only implicated in the women economic growth. The tourism development impacted the social lives of Dusun Ngentak women. The development of economy tourism system, the women groups, began to rise as a way to organized women who worked in the tourism sector in social activities, such as monthly "arisan" (regular social gathering). That activity meant to strengthen their social relations as the same economy actors.

The social harmony had formed by the group activity could support women's business/economy sustainability in Dusun Ngentak. Social connections which had been formed among the women can be social capital for them to develop better tourism in the future.

5 Conclusion

Basically, the HEPS installation had the two main purposes which also interconnected to each other. First, HEPS has the main task to guarantee the local needs of energy. In this case, hybrid energy was used as the electric power provider for people of Dusun Ngentak. Electricity was a technical aspect, in this case how it could be distributed to people to fulfill their needs on electricity. Second, the HEPS was expected to improve the economic condition of people nearby, through the usage of electricity as the support facilities of their economic activities and people empowerment in Dusun Ngentak.

By chasing the two purposes, the hybrid energy existence can intactly be innovative products which bring benefits to people. Thus, the role of hybrid energy will be able to reach the fair energy distribution for all layers and elements of people. In the case of women of Dusun Ngentak, we can see several opposite sides related to the changes in the process of energy distribution.

The dynamic of the changes in women of Dusun Ngentak can be seen in two phases, pre and post installation of HEPS. Before the installation, women were only active in the domestic sector, where women were busy with their household routine activities and worked in subsistent agriculture. They were positioned as the second actor in the social structure of Dusun Ngentak. Men domination was really powerful in all sectors, especially public sector which holds common interests. Men were the real actor who got the responsibility for the lives of family and society. While, women stayed to household activities such as babysitting children, cooking, or helping their husbands in the agriculture or looking after their cattle.

Women of Dusun Ngentak indirectly experienced injustice and limitation which occurred culturally. The perception which saw women as weak and placed them in the domestic sector was the form of the unfair social system. Consequently, a woman could not be a free and independent person, because the injustice which constructed socially led them to feel inferior. The social system which ran culturally was a sign that choices in the dominant or public sector would be fully managed men, although actually, women have the equal power to manage. Roles and status of women became unrespectable when they were in the public sector. In consequence, men will overpower the strategic sectors and involve the technology determination.

Since the development of Pantai Baru region, supported by the HEPS installation, there were many changes involved the role and status of women in public sector. Women who used to be dubbed as weak and incapable of doing things on public sectors, nowadays experiencing a paradigm shift. The development of Pantai Baru and installation of HEPS opened spaces for new activities to women so that they could express their capabilities and skills. Generally, electricity provision program and people economy development in Dusun Ngentak had given impact in upgrading women social status in society as well as their economic condition.

Women's position in public sector, particularly in the development process of Pantai Baru as tourism area had a significant change. Women have a great role in maintaining the tourism activities in Pantai Baru. Economy activities which were developed as a supplement in beach tourism had diffused into varied sectors, such as culinary entrepreneurs, souvenir merchants, even some jobs like keepers of a public toilet and playground rides. Activities which were dominated by women, soon be the major livelihood. The incomes earned helped the family economy, even so, they became the main income for their families.

Women of Dusun Ngentak did not only take part in culinary and tourism business. Actually, there were many other sectors where they participated and occupy the strategic positions. For example in the fishery, fishery development program lead women to take a strategic position as the main actors because it was targeted to empower women of Dusun Ngentak. Unfortunately, it did not work optimally because they could not operate it well. Only small numbers of people who can take benefit from it.

In general, women of Dusun Ngentak more concerned to develop culinary business and trading because they think those were more prospective and directly connected to the tourism sector of Pantai Baru. Culinary business and trading were the sectors which were handled by all women of Dusun Ngentak. The management was diverse, it could be self-managed as a family business, or was managed by a group of mothers who did not have the capital to start a business by themselves. So that, almost all of women in Dusun Ngentak involved in the culinary business and trading activities in Pantai Baru.

The condition was reversed compared to the one before the HEPS installation and tourism development. Women who were previously treated as the second actor, then become a central actor in the tourism development as well as their family economy. Since, without women's participation in Dusun Ngentak, tourism in Pantai Baru would never be developed until now. Women of Dusun Ngentak started to lift up their lives and found the balance in managing public sectors of Dusun Ngentak

Furthermore, the HEPS gave a chance for women to the wide access and was supported by their culinary businesses or trading activities. Although not all of the businesses in Pantai Baru used electricity power from HEPS, the even researcher found that in early 2016 some outlets had used electricity from the national electricity company (PLN). It happened because HEPS could not provide sufficient energy for the developing tourism necessity.

Although women were not directly involved in the installation process and management of hybrid technology in Pantai Baru, and the domestic sector was still women's main realm, at least there were changes in role and status of women. Their position became more respectable, and they have a prominent role in the technology and tourism development in Dusun Ngentak. This position emphasized that women as actors of renewable energy usage. They had the capability to do self-improvement through the usage of limited hybrid energy, then capable of upgrading their positions in the social structure of Dusun Ngentak.

Thank you for Amirrudin, Vicky Dian, and Fauzia Hisyam for contributing in the discussions and data submissions at the fieldwork.

References

[1] Amiruddin, et.al, 2012, "Menuju Keadilan Dan Kedaulatan Energi Terbarukan: PLTH sebagai Eskalator Transformasi Kehidupan Sosial Ekonomi Perempuan di Dusun Ngentak, Desa Poncosari, Kecamatan Srandakan, Kabupaten Bantul, Yogyakarta (Toward Justice and Renewable Energy Sovereignty: HEPS as the Escalator of Women's Social-Economic Transformation in Ngentak Poncosari, Srandakan, Bantul, Yogyakarta)". Report of Hibah Bina Desa, LPPM UGM. (in Bahasa Indonesia)

[2] Anggraeni Liestya et.al, 2010, "Pembangkit Listrik Tenaga Hibrid (PLTH) sebagai Penggerak Transformasi Sosial dan Kesejahteraan Komunitas Nelayan Pesisir Pantai Pandansimo, Dusun Ngentak, Desa Poncosari, Srandakan, Bantul, Yogyakarta (the Hybrid Electric Power Systems (HEPS) as the Social and Welfare Transformation Booster of Fishermen Community in the Coast of Pantai Pandansimo, Ngentak, Poncosari, Srandakan, Bantul, Yogyakarta)", Research Report of Hibah Mahasiswa, Departemen Sosiologi, Fisipol UGM. (in Bahasa Indonesia)

[3] Campbell, M.J, 1989, New Technology and Rural Development: The Social Impact, Routledge, London.

[4] Feenberg A, 2002, Transforming Technology: from Marxian to radical critique, Oxford University Press, New York.

[5] McIntyre A, 2008, Participation Action Research, Sage Publication, London.

[6] Scheer, Hermann, 2007, Energy Autonomy: The Economic, Social and Technological Case for Renewable Energy, Earthscan Publication.

[7] Widhyharto, D.S, et al., 2014, Penguatan Sosial-Ekonomi Kelompok Perempuan Pesisir: Konsep dan Praktik Pemanfaatan PLTH Pantai Baru, Yogyakarta (Social-Economical Strengthen Strategy of Coastal Women Groups in the Concept and Practice of the Usage of HEPS Pantai Baru), Departemen Sosiologi, Fisipol, UGM. (in Bahasa Indonesia)

[8] Widhyharto, D.S, 2015, "Perempuan Sebagai Agen Perubahan Sosial-Ekonomi Masyarakat Pesisir Sekitar PLTH, Kab.Bantul, DIY (Women as Agent of Social-Economical Changes of Coastal Society of HEPS installation area in Bantul, DIY)". Indonesian Journal of Community Engagement, Vol.1, No.1, Page156-171. (in Bahasa Indonesia)

Energy Monitoring System for Existing Buildings in Indonesia

Sentagi Sesotya Utami[1], Faridah[1], *Na'im* A. Azizi[1], *Erlin* Kencanawati[1], *M.* Akbar Tanjung[1], *Balza* Achmad[1]

[1]Department of Nuclear Engineering and Engineering Physics, Universitas Gadjah Mada, Yogyakarta, Indonesia

Abstract. Current studies conducted by JICA, AMPRI and IFC-World Bank, reported that large commercial buildings in Indonesia are not energy and water efficient. One of the cause is the lack of regulation. Meanwhile, effective regulations to reduce energy and water consumption are the concern mostly in a new building to obtain a building permit. This strategy is understandable as retrofitting existing buildings are often more difficult to be implemented, and enforcement is still a major issue in Indonesia. Local governments are currently working on streamlining building permit process as well as developing an online monitoring system for existing buildings. By applying a Building Energy Management System (BEMS) enables to reduce energy consumption up to 15%. An energy monitoring system was designed and installed through this research for Department of Nuclear Engineering and Engineering Physics (DNEEP) building, Faculty of Engineering, Universitas Gadjah Mada. It is a 20 years old two-story building used for educational activities, which consist of classrooms, laboratories, offices and storage spaces. An audit energy was done recently in 2015 where an energy consumption of 261.299,636 kWh/year.m2 was reported. In the existing condition, a power meter is inaccessible and therefore, the only feedback of occupancy behavior in the energy consumption is through the electricity bill. The previous study has shown that building occupants would behave more efficiently if the amount of energy used is notified, and the amount of energy savings are recorded. However, these energy monitoring systems are considered expensive and uniquely tailored for every building. This research aims to design and install a cost effective BEMS based on occupant's satisfactory assessment of the lighting, acoustics, and air conditioning quality. The data will be used as a decision supporting system (DSS) by building management through the use of a GUI. The design of the interface was based on a survey result from the prospective users. Installed energy monitoring system uses a current sensor with an accuracy of 98% and a precision of 0.04 A while the voltage sensor with an accuracy of 98% and a precision of 0.58 V. The performance testing shows that the number of web clients influences delay of data transmission. The result of the survey shows that GUI is categorized as fair in design without a significant difference between the perceptions of users with and without survey supervision. **Keyword— Building Energy Management System (BEMS); existing building; Decision Supporting System; sensor; GUI**

1. Introduction

Current studies conducted by JICA, AMPRI and IFC-World Bank, reported that large commercial buildings in Indonesia are not energy and water efficient. This is caused by ignorance from building owners and managers, architects and engineers, and lack of regulation. While regulations to obtain building permit currently is the most effective way of reducing energy and water consumption, these regulations are a concern mostly in new buildings and are implemented to obtain a building permit. This strategy is understandable as retrofitting existing buildings are often more difficult to be implemented, and enforcement is still a major issue in Indonesia. However, as the local governments of Jakarta, Bandung, Surabaya, and Makassar are currently working on streamlining building permit process as well as developing an online monitoring system for existing buildings, the next step is to regulate the techniques for energy and water efficiency improvements. Its implementation is far from ideal as there are no standardized monitoring system in most, if not all, existing buildings.

By applying a building energy monitoring system (BEMS) enables to reduce energy consumption up to 5-15% [1]. The building owner, building managers, and professionals will be able to operate and retrofit existing buildings to become more efficient in their energy consumption given the data obtained from the monitoring system. It is a significant benefit since the cost of energy is becoming more expensive, in line with the Government

 https://doi.org/10.1051/e3sconf/20184201003

policy to gradually lift the subsidy. The previous study has shown that building occupants would behave more efficiently if the amount of energy used is notified, and the amount of energy savings are recorded [2].

João Gil Josué et al. developed a BEMS for domestic buildings as a home energy management system [3]. Radio wave and microcontroller PIC18F2420 supported the data transmission system. The maximum line of sight between the data acquisition system and the data interface is 250 meters, sending graphics of voltage, current, power factor, active power and energy data.

The concept of the Internet of Things (IoT) was first used in 1999 by Kevin Ashton to the defined concept of computer and machines with sensors, which connect to the Internet to report status and accept control commands [4]. This concept provides the ability to develop an energy monitoring system supporting the complexity of a building usage. Web-based BEMS has already been developed using the internet from a Wi-Fi network as the data transmission system [5]. Since data are transmitted from the data acquisition system to the web server through the internet, the positioning of the data acquisition system and Wi-Fi router determine the data transmission quality indicated by time delay. A database is then provided at the end-user after the data is sent to the web server.

S. Tepić et al., developed a smart data transmission system for BEMS using an efficient software framework [6]. The efficient framework consisted of a message protocol based on HTTP application layer and JSON data package and Internet Protocol (IP) to increase flexibility in the use of physical layers that connect sensors with the entire wireless network. However, these BEMS are considered expensive and uniquely tailored for every building.

Existing buildings in Indonesia are mostly found without as-built drawing documents including information of the building electrical wiring and building service utilities. This is due to the lack of regulations and its reinforcement in the past, during the design, construction until the building commissioning phase. Therefore, to achieve the most effective BEMS implemented in existing buildings, a thorough investigation of the building condition and proper functionality of its premises, is required. As a consequence, the BEMS should be tailored to a unique sensor device and networks, data acquisition and transmission system, and customized the interface for data analysis by the BEMS user (building management) for each building.

This paper implemented the above idea to BEMS development in the Department of Nuclear Engineering and Engineering Physics (DNEEP) building, Faculty of Engineering, Universitas Gadjah Mada (UGM). It is a 20 years old two stories building that consists of classrooms, laboratories, offices and storage spaces. An audit energy was done recently in 2015 where an energy consumption of 261.299,636 kWh/year.m2 was reported. In the existing condition, a power meter is inaccessible and therefore, the only feedback of occupancy behavior in the energy consumption is through the electricity bill. This research aims to design and install an effective BEMS in the DNEEP building, based on occupant's satisfactory assessment of the lighting, acoustics, and air conditioning quality. The data will be used as a decision supporting system (DSS) by building management through the use of a GUI. The design of the interface was based on a survey result from the prospective users.

2. Method

There are mainly three important components in the BEMS, which leads to utilization of several steps and techniques during the development process (Fig. 1). Three components are the sensors, the data transmission system, and the interface as a DSS. It involves two functions, which are data acquisition and processing, and real-time monitoring based on Internet of Things (IoT).

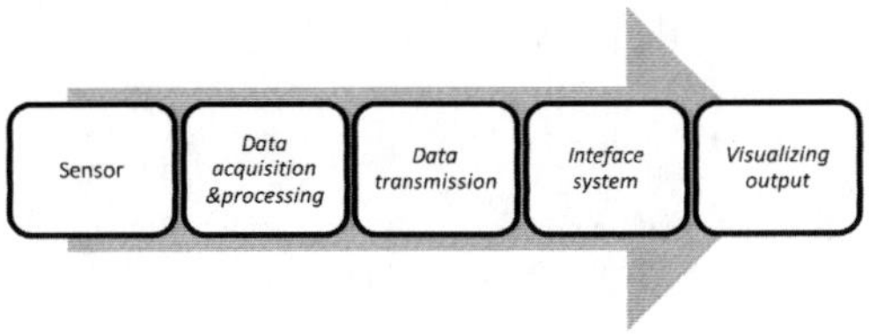

Fig. 1. Steps of the development process.

The system sensor is designed to measure the electricity load by using an electrical current sensor SCT 013-000 to sense the electrical current *i(t)* and sensing the voltage *v(t)* using sensor ZMPT101B. Sensors are tested using the input and output correlation used as the calibration factor. The correlation value depends on the condition of the electrical wires of the building's utilities. The calibration factor will be used by the microcontroller to process data and produce measured values, which are accurate and precise.

Using the measured *i(t)* and *v(t)*, the amount of energy consumption is generated by the BEMS based on the average power-usage. The electrical current and voltage passing through the system sensor are defined as (Fig. 2):

$$i(t) = I_m \cos(\omega t + \theta_i) \quad (1)$$

$$v(t) = V_m \cos(\omega t + \theta_v) \quad (2)$$

Meanwhile, the average power of one cycle with period *T* is given by:

$$P = \frac{1}{T}\int_0^T v(t)i(t)dt = V_{rms}I_{rms}\cos(\theta_V - \theta_i) \quad (3)$$

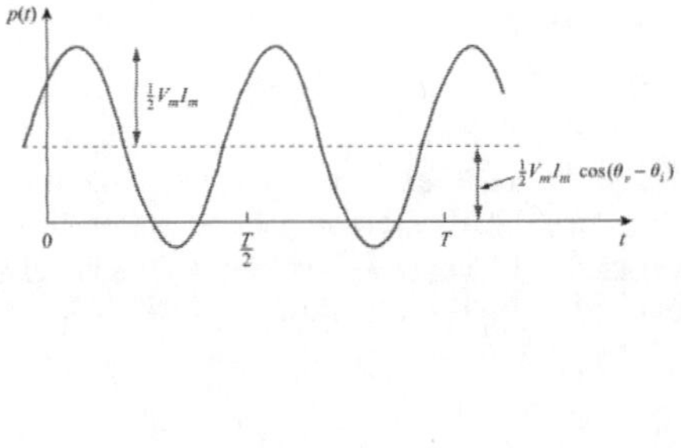

Fig. 2. Instantaneous power in circuits [7].

It seems as if $V_{rms}I_{rms}$ is already providing the power value needed. However, this value is not yet the real average power since there is still the factor of $\cos(\theta_V - \theta_i)$ from the Equation 3 above. This unreal average power is known as appearance power [8]. Ratio between the appearance power to the average power is known as the power factor shown in equation

$$pf = \frac{P}{V_{rms}I_{rms}} = \cos(\theta_V - \theta_i) \tag{4}$$

The energy used, W, that will be reported is calculated from the average power, P, being sampled for each period by Δt =1/50 s, or can be written as,

$$W = P.\Delta t \tag{5}$$

Output data from the sensors system are stored in the data logger, which these parts stand as the data acquisition system. Data can also be sent to a server using the IoT supporting system (Fig. 3). There are three characteristics of IoT, which are the perception layer, network layer, and application layer [9]. Perception layer is the layer that interacts with all the items outside the IoT system, which in this case are the sensors used to obtain the information. The network layer is also known as the transportation layer due to its function in transmitting data from the perception layer to the application layer. Shortly speaking, this layer has a function to transmit the data through the internet from microcontroller to the web server and from the web server to the client. Application layer plays a role in processing and analysing data obtained from the perception layer in the form of a GUI, which can be accessed by the users to get information about the energy consumption.

The network layer of the BEMS developed in this research refers to web application design with the RESTful Web Service system. It implements the architectural concept of Representational State Transfer (REST). Given the principle of RESTful Web Service, the data transmission protocol is using HTTP Protocol. Aside to the use of HTTP Protocol, this system also uses JSON as standard data format, which is understandable to the web client and web server. It also identifies each resource using Uniform Resources Identifier (URI) [10]. The performance of this network layer is indicated by the time delay.

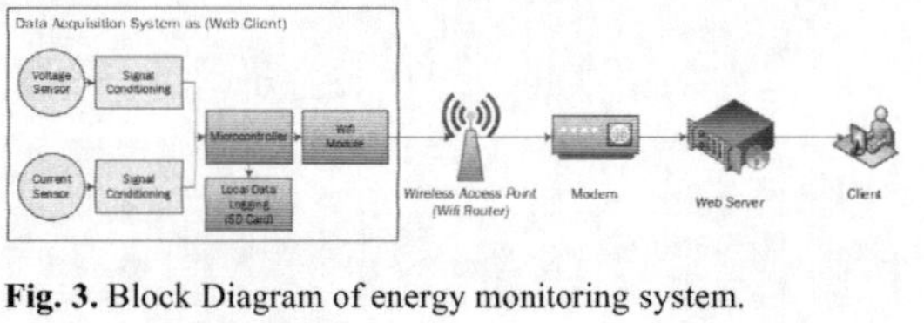

Fig. 3. Block Diagram of energy monitoring system.

The application layer refers to the design of Graphical User Interface (GUI) based on the concept of designing an Energy Monitoring and Targeting (EM&T). Slightly different with Building Energy Management System (BEMS) that provides a real-time control of the building utilities to perform its functions in terms of safety, efficiency, and comfort, EM&T only monitors the energy consumption by collecting data and analyzing the history. There are two main functions of EM&T. By monitoring the energy consumption and compare the energy usage of previous time or standards of typical buildings, the building user can control the energy use with EM&T. The second function is the ability to increase energy efficiency by defining an energy consumption target for the next period of analysis [11]. There are four monitoring processes in the BEMS that are considered during the GUI design, which are:

1. Data collection, a process in collecting data on the energy consumption of the area, monitored and also other relevant data.
2. Data Analysis, data received by the computer are examined of any errors and check whether the data are saved correctly for the analysis. The analysis is determined by the system requirement.
3. Reporting, a stage where the energy consumption report is produced and being compared with the target.
4. Action, implementation of the report by the building management. This action is needed to effectively reduce the energy consumption where the previous consumption was reported by the BEMS.

The client is able to monitor the energy consumption by accessing the web address using a common web browser. Every 15 minutes the electricity power usage, P and energy usage, W are sent from the sensors. An interface in the application layer is designed to provide easiness for the user in accessing information from the monitoring system. Data are presented in graphics and charts, which is then used as the statistical analysis in the energy consumption report. Several indoor quality parameters that are strongly related to the building energy consumptions, which are lighting, acoustics, and air conditioning quality, are also displayed and reported. This interface was design based on the clients need in order to create an interesting, informative and easy-used web interface. A survey was conducted to ensure the user's needs were already accommodated by the interface designed.

3. Results and Discussions

3.1 Sensors and Data Processing

3.3.1 Electricity Current Sensor System.

The sensor for electrical current is using SCT 013-000 with a microcontroller as a data processing tool shown in Figure 4.

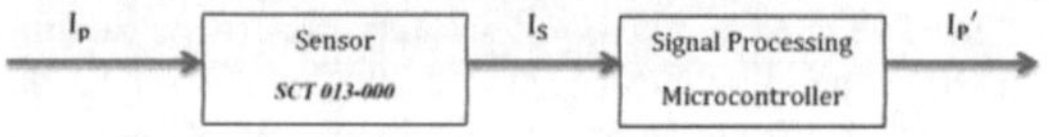

Fig. 4. The sensing method of the current sensor.

The SCT 013-000 sensor has a range of input AC current (I_p) of 0-100 ampere (A) and a range of output AC current (I_s) of 0-50mA. The transduction process uses the principle of induction as in an ordinary transformer. Electricity current measured that flows through the wire will cause magnetic field at the magnetic core and produced the output AC current (I_s) at the secondary coil. Zener diode in the internal sensor has a function as a circuit protector to avoid extremely high voltage due to the change in electrical current during transient conditions. Fig. 5 describes the relation between I_s and I_p or commonly known as a transfer function. The sensor transfer function is

$$I_S = 0.0005\, I_p \qquad (6)$$

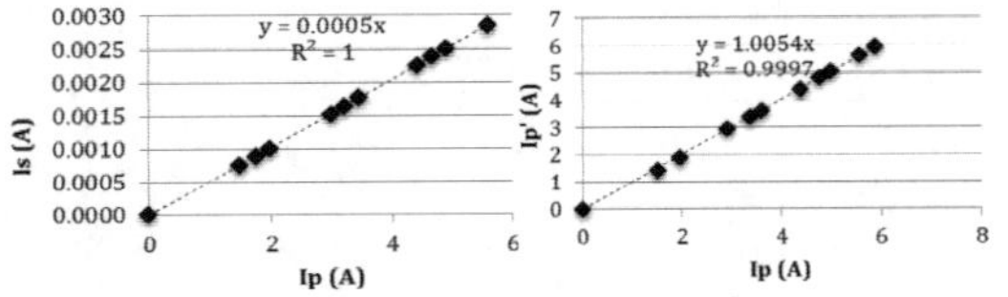

(a) (b)

Fig. 5. Transfer function of the sensor SCT-000, and (b) and with the entire system (current sensor and microcontroller)

This relation is then used as the calibration factor of the microcontroller Arduino with an output current measured, I_p' , which is similar to the actual electrical current. Meanwhile, Fig.6 describes how the system works, the relation between I_p and I_S. The accuracy and precession of the sensor system are 98 % and 0.04 A, respectively.

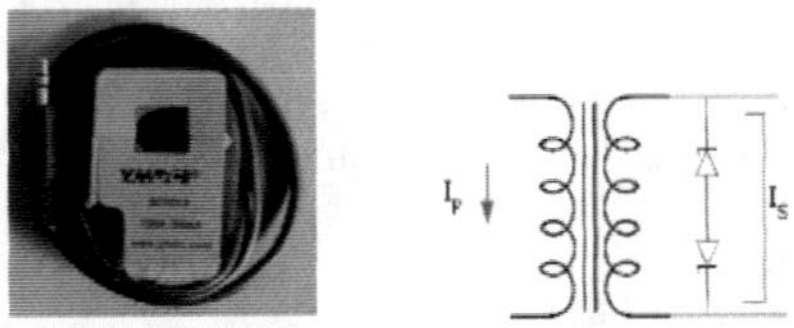

(a) (b)

Fig. 6. (a) Electricity current Sensor SCT-000, and (b) diagram of how the current sensor works

3.3.2 Electricity Voltage Sensor System.

The sensor system for measuring the electrical voltage used in this BEMS consists of a voltage sensor ZMPT101B and a microcontroller to process the data as described in the block diagram in Fig. 7.

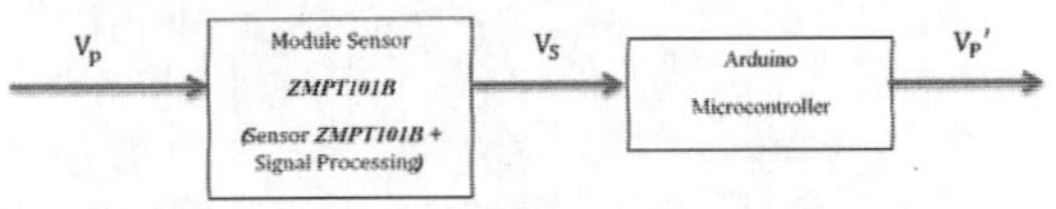

Fig. 7. The sensing method of the voltage sensor.

The ZMPT101B voltage sensor also works using the principle of a simple transformer. The measured or input voltage (V_p) will cause electric current (I_P) and induce the output electric current (I_S), which will flow with a resistance of R, creating the output voltage (V_S). Fig. 8 shows the relation between V_p and V_S. The transfer function of the voltage sensor is given by,

$$V_S = 0.0035\, V_p \qquad (7)$$

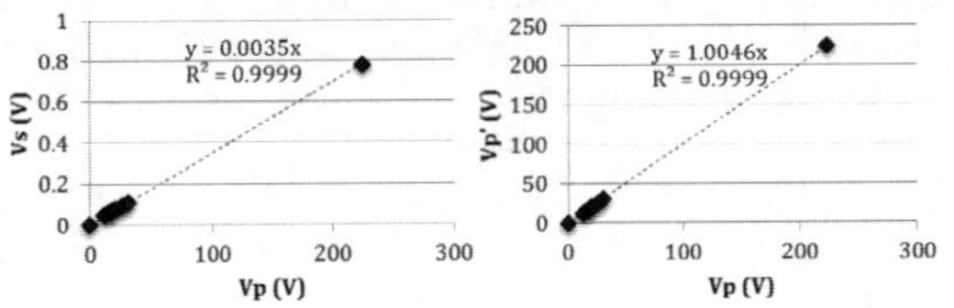

(a) (b)

Fig. 8. (a) Transfer function of the sensor ZMPT101B, and (b) and with the entire system (voltage sensor and microcontroller)

Similar to the current sensor, this transfer function is used as the calibration factor in the microcontroller to obtain the output voltage of the system V_p', which has the same value as the input voltage (V_p). Meanwhile, Fig. 9 describes how the system works and the relation between V_p' and V_p. The accuracy and precession of the sensor system are 98% and 0.58 V, respectively.

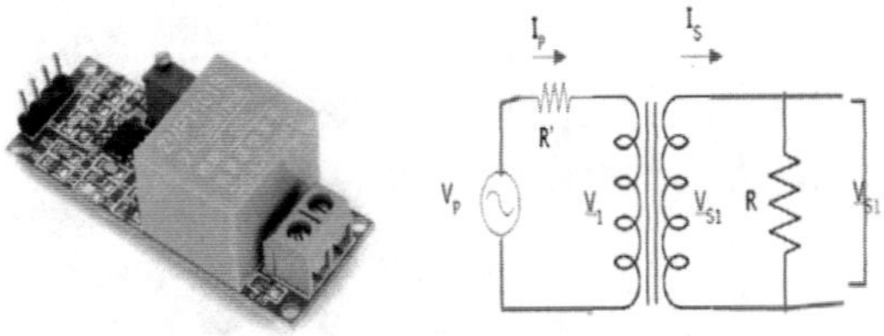

Fig. 9. (a) Electricity voltage sensor ZMPT101B, and (b) diagram of how the voltage sensor works.

3.3.3 Sensor positioning and installation

The positioning of the sensors is based on the analysis of the electrical wiring diagram of DTNTF building. As in the majority of an existing building in Indonesia, the as-built drawing of the case study also does not exist. Therefore no electrical wiring diagram is available. An on-site survey is required where every wire is traced to get the electrical wiring diagram of the electrical lighting and AC of DTNTF building. Ideally, the wiring of each building equipment or utilities should be traceable in order to have the energy consumption data of this equipment. If this information is available, the building manager will be able to come up with strategies for reducing the energy consumption. The closest approach in tracing the equipment's wiring is by grouping equipment into zones, which is the first attempt of this BEMS designed.

3.2 Data Transmission

Beside as a data processor in the sensors system, the microcontroller Arduino also acts as a web client. As a web client, Arduino sends data from the sensor to the web server. The data consists of V_{rms}, I_{rms}, power factor, average power, and the energy consumption. The complete sensor system with the Arduino web client is a box sensor. Each box sensor consists of four packages of data sensor and therefore enables to measure four different electricity wires. The data transmission algorithm within the Arduino is using HTTP method followed by URL and data packet. This data packet consists of measured data, sensor identifier, and zone identifier. All the data within the Arduino are sent periodically at certain intervals. In this BEMS, data are sent every 15 minutes.

Web server is used to save the web application at a certain domain (Fig. 10). The web server in the web application is using the RESTful Web Service system, which creates HTTP Router. These routes were written in Hypertext Preprocessor (PHP) coding language and can be accessed by the web client by using URL and HTTP request method. Every HTTP Router has a different response and function. With this HTTP Router, the web client is able to send data to the database and access the resources that are kept in the web server or database system.

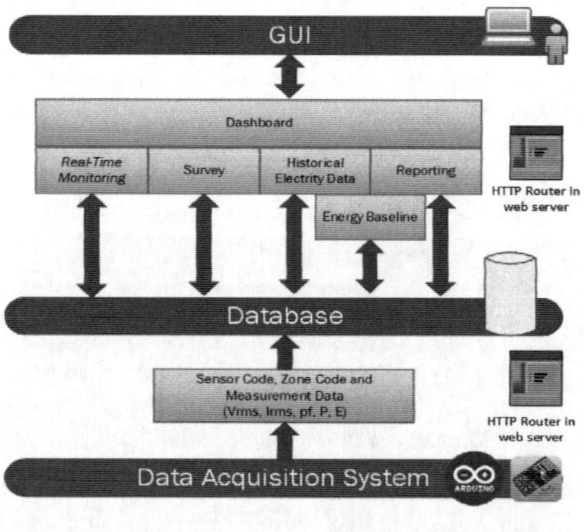

Fig. 10. Structure and function design of HTTP Router in *web server*.

The figure illustrates the HTTP Router in the web server of the BEMS. The main function of the HTTP Router can be categorized as two paths, which are the path from GUI and the path from the microcontroller Arduino as the web client. In the GUI path, there is an HTTP Router that can be used by the user to access pages of the dashboard as the home page of the BEMS GUI. From the dashboard pages, users can access real-time monitoring pages, survey, historical data and report of the energy consumption. In each page, the HTTP Router is also being functioned to process and display the data from the database as what is needed for each page. In the Arduino path as the web client, there is an HTTP Router that serves to receive the data measured being sent by the Arduino and save the data into the database.

The result of testing the time delay of data transmission system by varying the number of web client is shown in Fig. 11. Web clients used in the testing of the data transmission system are Arduino and users who access GUI by using a web browser. The differences between those web clients are Arduino as web clients send data to a web server however users request data from the web server to display the web page. The average delay for each variation of web clients is 17 ms for1 web client, 64 ms for 10 web clients, 253 ms for 50 web clients and 566 for 100 web clients. Based on the result, energy monitoring system can be developed on a larger scale with the consequence of long time delay.

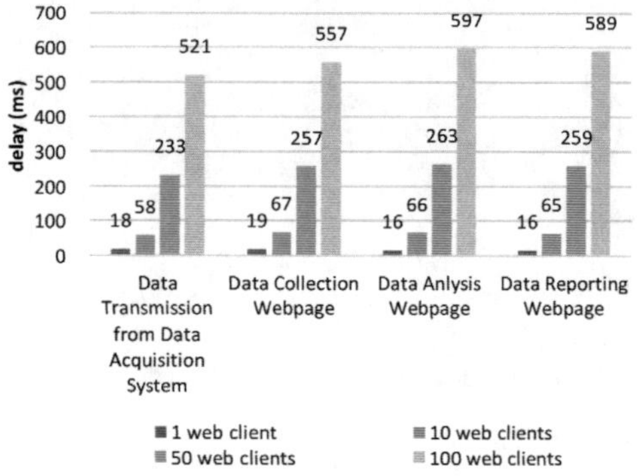

Fig. 11. Performance testing of energy monitoring system by varying the number of web clients.

3.3 Interface

The concept of EM&T is used to develop the GUI of this system as mentioned earlier. The dashboard of this BEMS's GUI is shown in Fig. 12. This version is still in Bahasa Indonesia. The monitoring system is part of the BEMS, which integrates with the monitoring of the lighting condition, temperature, humidity, and volume of CO2 of the building.

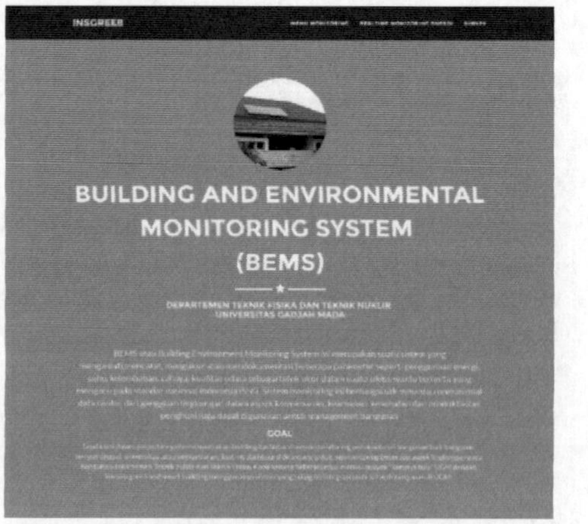

Fig. 12. Overview of BEMS, navigation of the GUI and the location of the installation of the monitoring system.

3.3.1 Data Collection

A Real-time monitoring inside the dashboard is display as graphics that show the energy consumption in real time starting from 00.00 am each day as shown in Fig. 13. Data of real-time are within the interval of 15 minutes complying with the time interval of data transmission from the Arduino, web client.

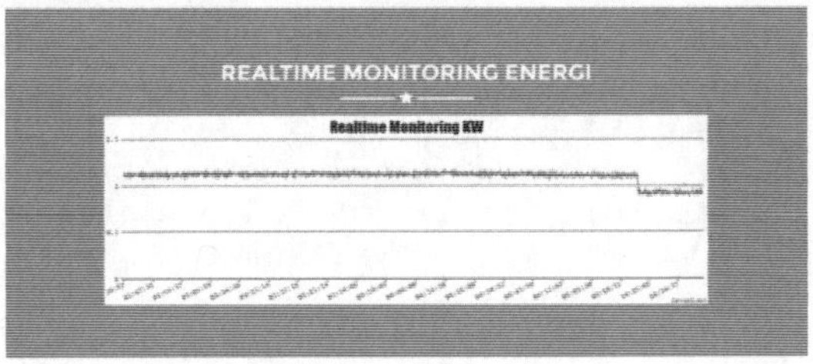

Fig. 13. Chart of real-time energy monitoring.

3.3.2 Data Analysis

In the EM&T concept, data are display in pages of data history to meet the user needs, which is the building management. Data history enables to monitor the daily energy consumption and compares with the baseline condition (see Fig. 14). Energy baseline is the energy usage threshold defined by the building management. The threshold of power usage from the energy baseline from time to time is known as the power baseline. Both data of the energy and power baseline are provided through audit energy of the existing condition.

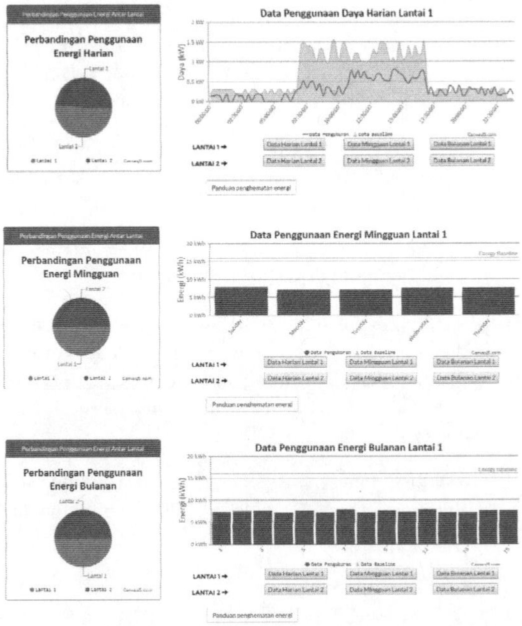

Fig. 14. Chart of energy usage analysis with energy baseline.

Under the graphic that displays the energy consumption, some energy-saving strategies are provided in the form of a slideshow, in Bahasa Indonesia, to help the user in understanding technical and non-technical energy-saving strategies (see Fig. 15).

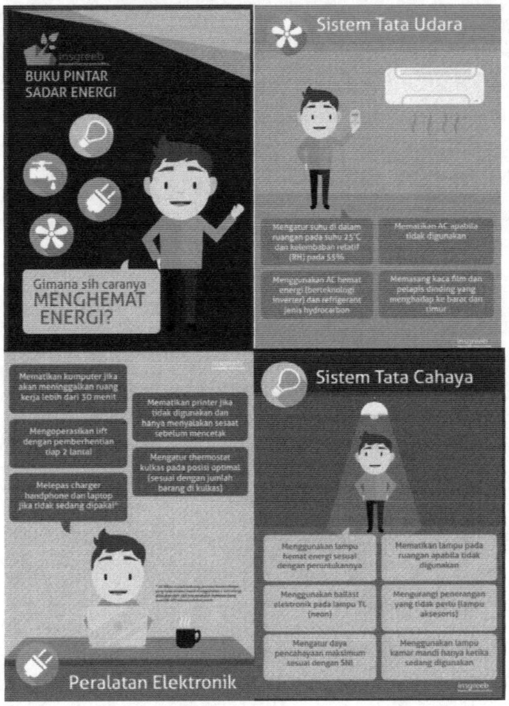

Fig. 15. Energy saving tips to help client carry out energy management (in Bahasa Indonesia).

3.3.3 Data Reporting

Annual report of the energy consumption is also provided in the GUI of the energy monitoring system (see Fig. 16). The annual report is chosen since the energy use cycle is for a year. Energy consumption is influenced by the surrounding climate of the building and activities of its surrounding, which is usually observed as one year cycle.

The final step of the EM&T concept is the action which is not yet implemented in this BEMS since it depends on to the building management policy given the energy consumption data. The use of GUI based on the survey conducted by building user has shown that the design of the GUI is categorized as fair, with a score of 3.71 points out of 5 for respondents that have gone through some supervision before filling out the survey. There is no significant difference in the GUI perception of two different respondents group, with and without survey supervision.

 https://doi.org/10.1051/e3sconf/20184201003

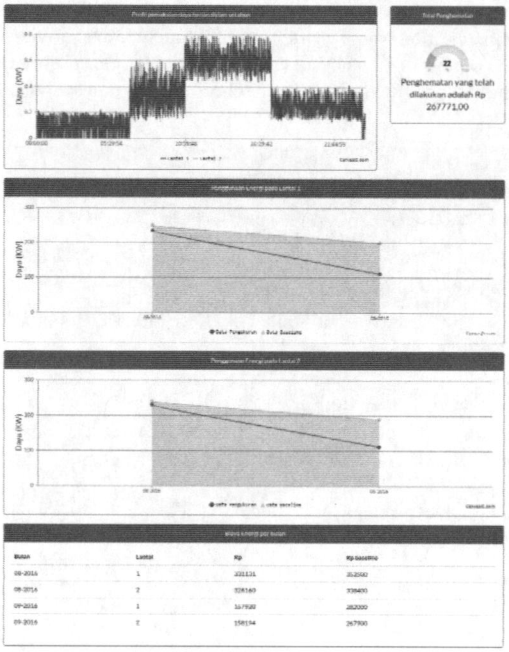

Fig. 16. Annual reporting of energy monitoring.

4. Conclusion

This research has shown that a building energy management system (BEMS) for every existing building is uniquely designed and specifically tailored due to the complexity of the existing wiring system, building zonation, and building utility or equipment that are being operated. It requires several steps from the characterization of the sensors, reliability test of the data transmission system and a survey to develop a human-friendly interface as an efficient DSS for the building management.

Based on test results, the current sensor has an accuracy of 98% and a precision of 0.04 A while the voltage sensor has an accuracy of 98% and a precision of 0.58 V. The average delay of data transmission testing for each variation of web clients are 17 ms for one web client, 64 ms for 10 web clients, 253 ms for 50 web clients and 566 for 100 web clients. Based on the result data transmission testing, energy monitoring system can be developed on a larger scale with the consequences of long time delay.The results of a user survey show that the design of the GUI is categorized as fair, without a significant difference between the perceptions of users with and without survey supervision.

The results also provide important information for further improvement of the system as more sensors will be implemented for the entire building and can be used as a benchmark for other similar building. In other words, the final product presented in this paper is a bone of a BEMS for the DTNTF building. The target of the building energy efficiency can be obtained after several years of data collection. Approximately 10% of the energy consumptions for several zones in the building area is provided as a baseline, and the data complies with the audit energy data gathered in 2015.

References

[1] S. Darby and others, The effectiveness of feedback on energy consumption, Rev. DEFRA Lit. Metering Billing Direct Disp 486, 2006, (2006)

[2] W. Kempton, J.M. Darley, P.C. Stern, Am. Psychol 47, 1213-1223 (1992)

[3] J.G. Josue, J.M. Pina, M.V. Neves, Doctoral Conference on Computing Electrical and Industrial Systems 2011, 437-444 (2011)

[4] D. Norris, The Internet of Things: Do-It-Yourself at Home Projects for Arduino, Raspberry Pi and BeagleBone Black (McGraw-Hill Education, New York, 2015)

[5] Y. Qu, H. Wang, S.M. Lund, H.D. Chiang, T. Wang, Design and implementation of a Web-based Energy Management Application for smart buildings, 2013 IEEE Electrical Power Energy Conference (EPEC), 1-6 (2013)

[6] S. Tepić, P. Pejić, J. Domšić, H. Mihaldinec, H. Džapo, 2015 38th International Convention on Information and Communication Technology, Electronics and Microelectronics (MIPRO), 143-148 (2015)

[7] C. Alexander, M. Sadiku, Fundamentals of Electric Circuit Analysis (McGraw-Hill Higher Education, Columbus, 2012)

[8] W.H. Hayt, J.E. Kemmerly, Engineering Circuit Analysis (Mc Graw Hill Higher Education, Columbus, 1971)

[9] X. Yang, Z. Li, Z. Geng, H. Zhang, Internet of Things, 388-393 (Springer, Berlin, 2012)

[10] S. Abeysinghe, Restful PHP web services: Learn the Basic Architectural Concepts and Steps Through Examples of Consuming and Creating Restful Web Services in PHP (Packt Publishing, Birmingham, 2008)

[11] P. Warburton, K. J. Butcher, Chartered Institution of Building Services Engineers, Eds, Building control system, 2. ed, London: CIBSE, (2009)

Mass and Energy Balance Analysis of Methanol Production Using Atmospheric CO2 Capture with Energy Source from PCMSR

Andang W. Harto[1] and *Mella* Soelanda

[1]Department of Nuclear Engineering and Engineering Physics, Faculty of Engineering, Universitas Gadjah Mada, Indonesia
andangftn@yahoo.com, [1]andangwh@ugm.ac.id

Abstract. The rising of atmospheric CO2 concentration is the major source to global warming system. Many methods have been proposed to mitigate global warming, such as carbon penalty, carbon trading, CO2 sequestration, etc. However these proposed methods are usually uneconomical, i.e., these methods do not produce economic valuable substances. This paper will propose a method to absorb atmospheric CO2 to produce economic valuable substances such as methanol, dimethyl ether, ethylene, several hydrocarbon substances and derivatives and several graphite substances. This paper is focused on methanol production using atmospheric CO2 capture. The overall process is endothermic. Thus a sufficient energy source is needed. To avoid more CO2 emission, the energy source must not use conventional fuels. To assure the continuity of energy deliberation, nuclear energy will be used as the energy source of the process. In this paper, the Passive Compact Molten Salt Reactor (PCMSR) will be used as the energy source. The 460 MWth PCMSR is coupled with atmospheric CO2 capture, desalination, hydrogen production and methanol production facilities. The capturing CO2 capacity is 7.2 ton/h of atmospheric CO2. The valuable outputs of this system are 3.34 ton/h of H2, 34.56 ton/h of O2, 5.24 ton/h of methanol and 86.74 MWe of excess electricity. **Keywords—CO_2; methanol; economic product; PCMSR**

1 Introduction

The increasing of atmospheric CO2 concentration becomes the most important environmental issue recently. The increase of CO2 in the earth atmosphere gives the effect of global warming which in turn causes global climate change [1]. The global climate change will cause many serious problems to human civilization. The examples of serious effects global climate are several natural disasters such as flooding, drought, harvesting failure, etc. [2]. Several actions have been proposed, and some of them have been implemented. The common goal of these actions is to reduce human-made CO2 emission and to increase CO2 absorption of both naturally or human-made. Some of the examples of actions to reduce CO2 emission are carbon penalty to every entity who emit CO2 by using conventional energy resources or other emitting CO2 process, carbon certificate and carbon trading for every entity who able to substitute the emitting CO2 processes with the other processes that emit less CO2 [3]. The examples of actions to increase CO2 absorption or to limit the decreasing of CO2 absorption are deforestation penalty and CO2 sequestration.

Most of these proposed methods do not produce real economics products, so the implementation of these actions will be viewed as an economic burden of human civilization. The consequence, massive law enforcement is required to guarantee these actions to be a success. It is interesting to invent a method of atmospheric CO2 increasing mitigation that able to produce real economics valuable products. This will promote the global warming mitigation by natural human behavior.

2 Basic Theory

2.1 Valuable substances from CO_2

The CO2 can be reacted with H2 to produce methanol (CH3OH) [4]. The methanol itself is a valuable substance. The methanol also can be processed to produce more valuable substance. For example, the methanol can be dehydrated to produce dimethyl ether (DME) [5]. Dimethyl ether can be dehydrated to produce ethylene [6]. Ethylene can be polymerized to produce several synthetic hydrocarbon substances. All of these substances are economically valuable and also feedstock to produce more valuable substances. Thus CO2 is actually can be processed to produce several economically valuable substances.

In recent technology industrial scale, CO2 is produced form other economically valuable substances. Production cost consideration and limiting the availability these substances limit the amount of CO2 production, and thus limit the number of several valuable substances produced from CO2. If it can be captured economically, the atmospheric CO2 will become huge and cheap CO2 resources to produce several economically valuable materials. The consequence, by this method the mitigation of global warming will boost economic development and will be no longer viewed as an economic burden for human civilization.

2.2 Atmospheric CO_2 Capture

There are two types CO_2 capture method. The first is CO_2 capture from concentrated CO_2 sources such as exhaust stack of coal or oil-fueled power plants or engines. The types of methods are relatively well established. However, these methods are not suitable to capture CO_2 from diffuse sources such as vehicles or other diffuse sources in which emit far more CO_2 than concentrated sources around the world. Thus the method to capture CO_2 directly from earth atmosphere is far more interesting.

Stolarrof et al. propose a method to capture CO_2 directly from earth atmosphere [7]. CO_2 will not be utilized after capturing but will be sequestrated. However, the Stolarrof method is interesting due to the ability to capture atmospheric CO_2 is a simple manner. This method will be referred I this paper.

In the Stolarrof method, atmospheric air is contacted with NaOH solution, and the CO_2 will be captured by this following reaction [7] :

$$2NaOH_{(aq)} + CO_{2(g)} \Rightarrow Na_2CO_{3(aq)} + H_2O_{(l)} \quad (1)$$

The Na_2CO is then regenerated by this following reaction [7] :

$$Na_2CO_{3(aq)} + Ca(OH)_{2(s)} \Rightarrow 2\ NaOH_{(aq)} + CaCO_{3(s)} \quad (2)$$

Thus the CO_2 is trapped in solid $CaCO_3$. In the original Stolarrof proposal, the $CaCO_3$ is directly sequestrated. This means the overall process is costly and unsustainable because need $Ca(OH)_2$ continually.

In this paper, the $CaCO_3$ is regenerated by this following reaction :

$$CaCO_{3(s)} \Rightarrow CaO_{(s)} + CO_{2(g)} \quad (3)$$

3

The CaO is then reacted with water to obtain $Ca(OH)_2$ with the following reaction :

$$CaO_{(s)} + H_2O_{(l)} \Rightarrow Ca(OH_2)_{(s)} \quad (4)$$

The overall process will be sustainable in term of NaOH and $Ca(OH)_2$. However, it is useless if the $CaCO_3$ regeneration is done in an atmospheric condition, so the CO_2 is released back into the atmosphere. In this paper, the $CaCO_3$ regeneration must be done in a reactor which supplied with hydrogen so the released CO_2 will react with hydrogen to produce methanol by this following reaction [4] :

$$CO_{2(g)} + 3\ H_{2(g)} \Rightarrow CH_3OH_{(g)} + H_2O_{(g)} \quad (5)$$

The overall system of atmospheric CO_2 capture followed by methanol synthesis, methanol dehydration, DME dehydration and synthetic hydrocarbon synthesis is shown in Fig. 1.

2.3 Hydrogen Production

According to equation (4), hydrogen is needed to convert CO_2 to methanol. Free hydrogen is unavailable naturally in a sufficient amount. Thus free hydrogen must be produced from naturally abundant feed materials. Hydrogen can be produced from conventional energy resources by using steam reforming method [8]. However, this method emits CO_2 thus this method is not suitable to the ultimate goal of mitigation of global warming by reducing atmospheric CO_2 concentration.

Hydrogen can be produced by using water as feedstock with the following reaction :

$$H_2O_{(l)} \Rightarrow H_{2(g)} + 0.5\ O_{2(g)} \quad (6)$$

The production of hydrogen from water does not emit CO_2. There are several methods to produce hydrogen from water. The high-temperature steam electrolysis (HTSE) using solid oxide electrolyze cell is the efficient and low-cost method for producing hydrogen from water [9][10]. This method is used as a reference method in this paper. Fig. 2 shows the schematic diagram of solid oxide electrolyzer cell.

2.4 Energy Consumption

The overall reaction of capturing CO_2 to become the final product of methanol is endothermic, means input energy is needed.

The reaction (1) is exothermic with Δh_0 = -109.4 kJ/mole of CO_2 and Δg_0= -56.1 kJ/mole of CO_2. With the negative value of Δg_0, the reaction (1) is spontaneous. The reaction (2) is exothermic with Δh_0 = -5.3 kJ/mole of Na_2CO_3 and Δg_0= -18.2 kJ/mole of Na_2CO_3. Also, the reaction (2) is spontaneous.

The reaction (3) is endothermic with Δh_0=177.99 kJ/mole of $CaCO_2$. The reaction (4) is exothermic with Δh_0= -65.3 kJ/mole of CaO. The reaction (5) or methanol synthetic reaction is exothermic with Δh_0 = -102.42 kJ/mole of CO_2.

The reaction (6) or hydrogen production is the most energy consume reaction. Fig. 3 the energy needed to produce hydrogen from water per mole of hydrogen [11].

To produce 1 mole of methanol, 3 moles of hydrogen must be produced, 1 mole of CO_2 must be captured, 1 mole of Na_2CO_3, of $CaCO_3$ and of CaO must be regenerated. Thus the energy needed per mole methanol production in ideal condition is (3×250 - 109.4 - 5.3 + 177.99 - 65.3 - 102.42) kJ/mole = 645.79 kJ/mole. The energy needed in real condition is higher due to energy losses and the fact that not all of energy released by the exothermic reactions are recoverable.

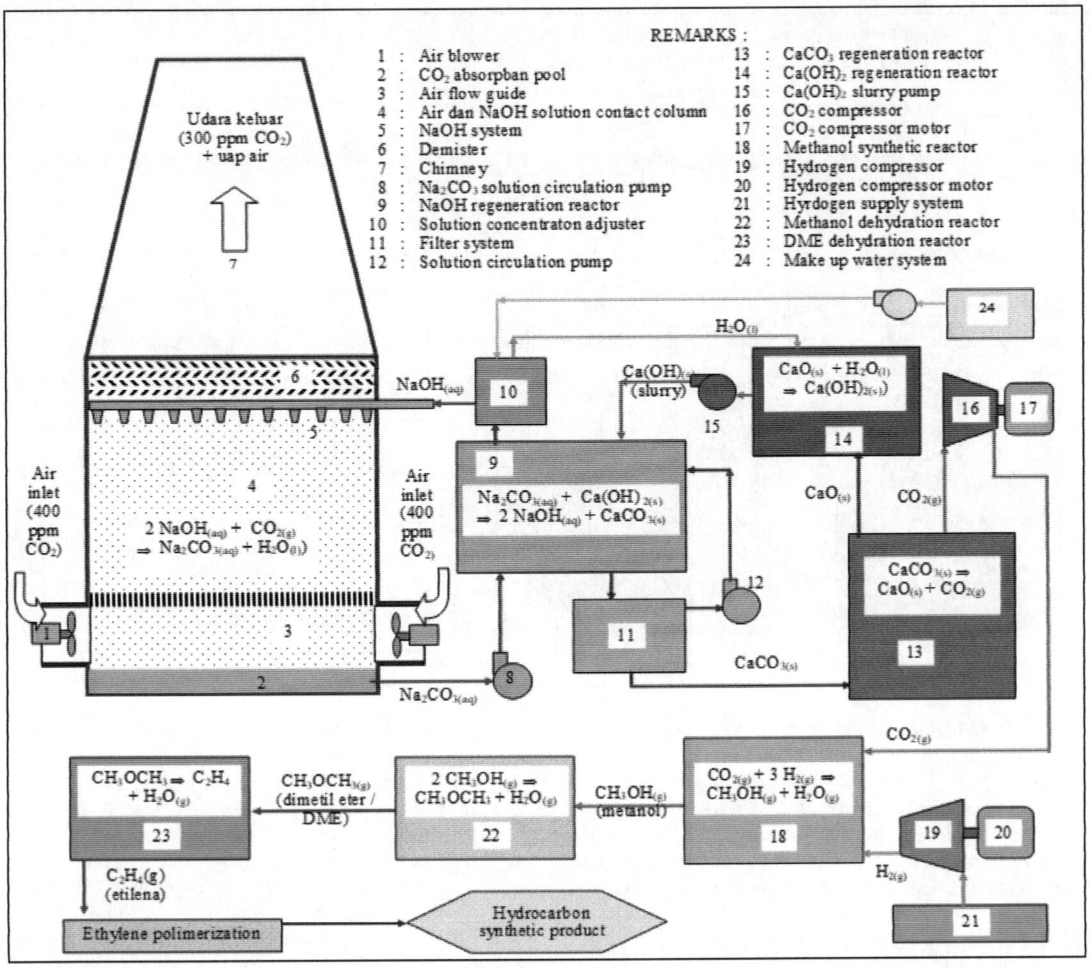

Fig. 1. The overall system of atmospheric CO_2 capture followed by methanol synthesis, methanol dehydration, DME dehydration and synthetic hydrocarbon synthesis.

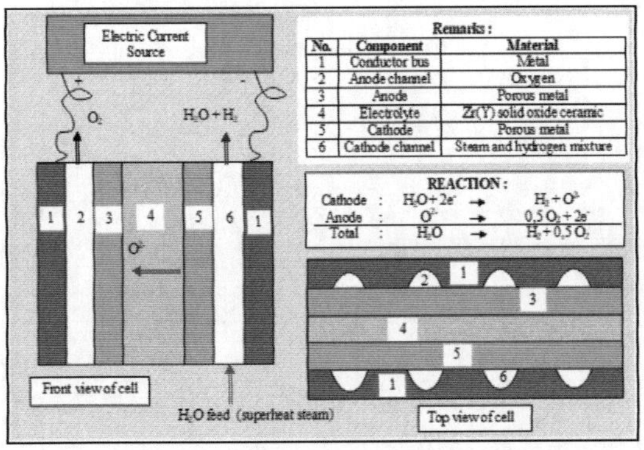

Fig. 2. Schematic diagram of Solid Oxide Electrolyze Cell

2.5 Energy Source

The fact that the overall reaction is endothermic means that an energy source is needed. The fossil energy sources must be excluded because of the emitted CO_2 and thus contra productive with the ultimate goal to mitigate global warming by absorbing CO_2.

Due to the fact that the overall process is extensive energy consume, the massive and continuous energy sources is needed. Thus nuclear energy source is the most suitable for this system.

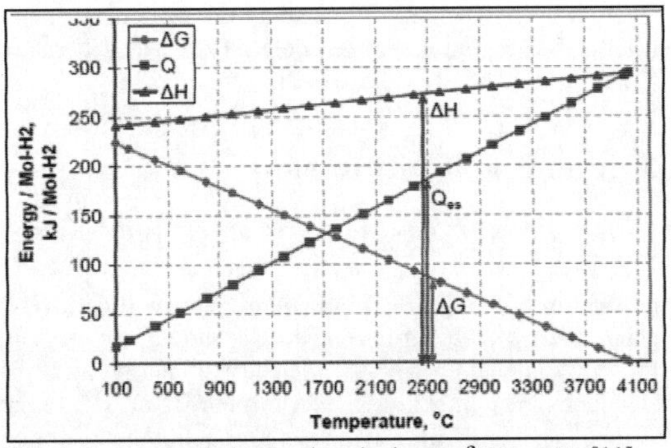

Fig. 3. Energy needed to produce hydrogen from water [11]

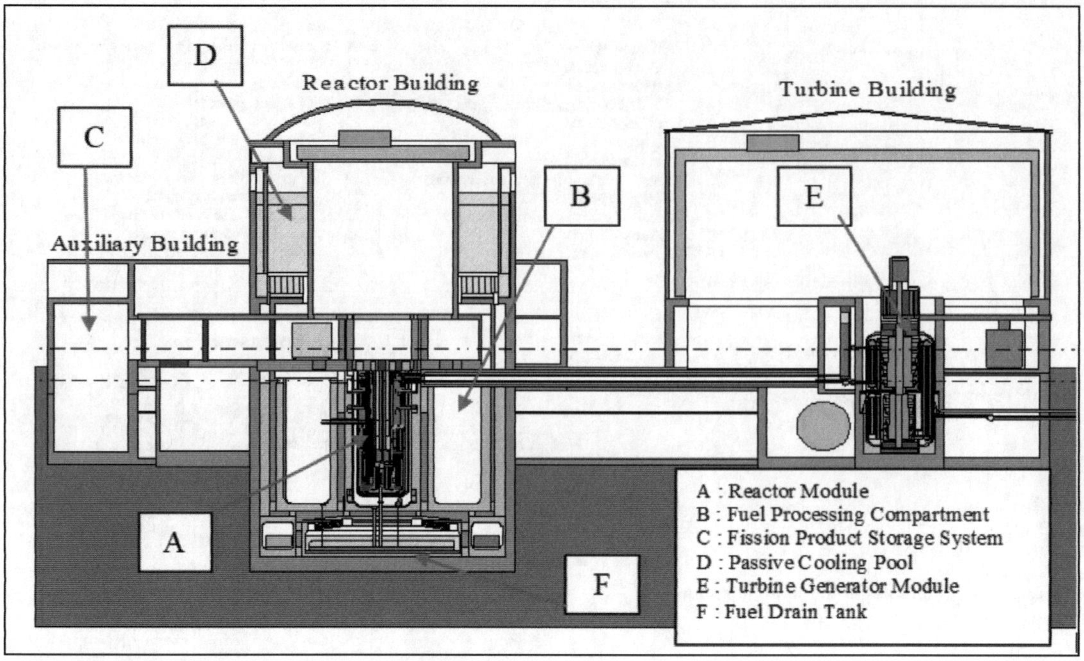

Fig. 4. The schematic diagram of PCMSR [12]

Regarding the consideration of long-term nuclear resources sustainability, the recent nuclear reactor technology that uses U-235 as fissile material is not recommended due to the natural U-235 resources scarcity.

The advanced breeder reactor utilizes more abundant fertile natural resources is more suitable. In this paper, the PCMSR (Passive Compact Molten Salt Reactor) which utilizes thorium nuclear fuel resources is applied to the energy source of the capturing CO2 to produce methanol process.

2.6 PCMSR (Passive Compact Molten Salt Reactor)

PCMSR is a variant of molten salt reactor that uses liquid fluoride salt mixture of 70 % mole LiF, 29 % mole ThF_4 and 1 % mole UF_4. The thorium is converted to fissile U-233 which undergoes fission reaction and produces the reactor power. The PCMSR use graphite as moderator and LiF-NaF-KF as intermediate coolant [12]. The schematic diagram of PCMSR is shown in Fig. 4.

3 Result and Discussion

1) Energy balance of hydrogen production

Schematic diagram of energy balance of hydrogen production using High Temperature Steam Electrolysis using energy source from PCMSR is shown in Fig. 5. In Fig. 5, Q is heat energy, W is electrical energy, and η is efficiency. The index of R refer to the reactor, HE refers to a heat exchanger for hydrogen production heat supply, TG refers to a turbine generator, AC/DC refer to the electrical supply system for electrolysis process, P refers to pump. In this analysis, it assumed that η_{HE} = 0.99, $\eta_{electrolizer}$ = 0.98, $\eta_{AC/DC}$ = 0.97, η_{TG} = 0.53, η_P = 0.70 and η_{Blower} = 0.70.

The thermal power of PCMSR is assumed to be 460 MWth. The hydrogen production capacity is set as 1.2 kg/s (4.32 ton/h) or 0.6 kmole/s = 600 mole/s. The HTES is assumed to be operated at 1000 °C. According to Fig. 3, Δg = 171 kJ/mole of H_2 and Δh = 251 kJ/mole of H_2 and thus $T\Delta s = \Delta h - \Delta g$ = (251 – 171) kJ/mole = 80 kJ/mole of H_2.

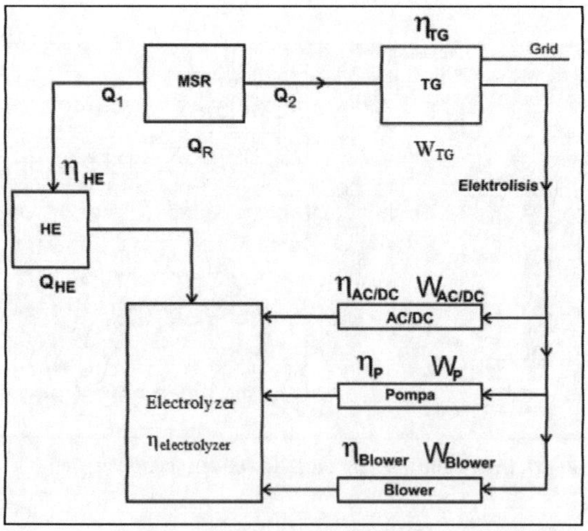

Fig. 5. Schematic diagram of energy balance of hydrogen production using High-Temperature Steam Electrolysis using energy source from PCMSR

The Δg component must be supplied as electrical energy. Thus the electrical power needed to supply the Δg component is

$$W_{EG} = \frac{\dot{m}_{H_2}\Delta g}{\eta_{electrolizer}} = \frac{600\,(\text{mole/s})\times 171\,(\text{kJ/mole})}{0.98} = 104\,\text{MWe}$$

The electrical energy also must be supplied to feed heater pump, recirculation water pump and gas mixture blower. The calculated overall water pumping power is small, i.e., 0.4 kWe and the calculated blower power is 2 MWe. Thus the total electrical power to be supplied for overall electrolyze process is

$$W_E = \frac{W_{EG}}{\eta_{AC/DC}} + \frac{W_P}{\eta_P} + \frac{W_{blower}}{\eta_{blower}} = \left(\frac{104}{0{,}97} + \frac{0.0004}{0.7} + \frac{2}{0.7}\right)\text{MWe}$$
$$= 110.36\,\text{MWe}$$

The $T\Delta s$ component must be supplied as heat energy from the reactor. The electrical energy loss in electrolyzer will become heat energy, and thus the thermal power to be supplied to reactor is

$$Q_{EG} = \frac{\dot{m}_{H_2}T\Delta s}{\eta_{HE}} - (1-\eta_{electrolizer})W_{EG}$$
$$= \frac{600\,(\text{mole/s})\times 80\,(\text{kJ/mole})}{0.99} - (1-0.98)\times 104\,\text{MWth}$$
$$= 48.48\,\text{MWe} - 2.1\,\text{MWth} = 46.38\,\text{MWth}$$

Heat also must be supplied to evaporate the feed water. To produce 0.6 kmole/s of hydrogen, 0.6 kmole/s or 10.8 kg/s of water is needed. The heat of evaporation of water is 2400 kJ/kg of water. Thus the evaporation heat needed to produce 0.6 kmole/s of hydrogen is

$$Q_{evap} = \frac{\dot{m}_{H_2O}\Delta h_{evap}}{\eta_{HE}} = \frac{10.8\,(\text{kg/s})\times 2400\,(\text{kJ/kg})}{0.99} = 26.18\,\text{MWth}$$

The total thermal power to be supplied by reactor to produce 0.6 kmole/s of hydrogen is

$$Q_1 = Q_{EG} + Q_{evap} = 46.38\,\text{MWth} + 26.18\,\text{MWth} = 72.56\,\text{MWth}$$

The PCMSR has the thermal power Q_R = 460 MWth and the overall turbine-generator efficiency of 0.53. A part of this thermal power must be supplied to the hydrogen production system, i.e., Q_1 = 72.56 MWth. Thus the thermal power supplied to the turbine system is

$$Q_2 = Q_R + Q_1 = 460\,\text{MWth} + 72.56\,\text{MWth} = 387.44\,\text{MWth}$$

The electrical output of turbine system is therefore

$$W_{TG} = \eta_{TG}Q_2 = (0.53\times 387.44)\ \text{MWe} = 205.34\,\text{MWe}$$

A part of the turbine electrical output power must be supplied to the hydrogen production system, i.e. $W_E = 110.36\,\text{MWe}$. Thus the turbine still produces extra electrical power of

$$W_{EX} = W_{TG} - W_E = (205.34 - 110.36)\,\text{MWe} = 94.98\,\text{MWe}$$

2) *Energy balance of atmospheric CO_2 capture*

The atmospheric CO_2 capture system is assumed to be designed capture 2 kg/s (7.2 ton/h) or 45.45 moles/s of CO_2. The ambient air inlet is assumed to contain 400 ppm of CO_2. The capture system is assumed able to reduce the CO_2 content to become 300 ppm as the air flows out from the capture system. This means the mass flow rate of air is 20 ton/s.

To capture 45.45 mole/s of CO_2, 45.45 moles/s of Na_2CO_3, $CaCO_3$ and CaO must be regenerated. The energy released from the exothermic reaction of the capturing CO_2 cycle is assumed to be unrecoverable. Thus the energy needed for capturing CO_2 process cycle is equal to the energy needed for the endothermic reaction of this cycle, i.e., reaction (3). The energy is supplied in the form of heat energy. It is assumed that the heat energy is supplied via an electrical heater, which the electricity is supplied by the extra electrical power of the PCMSR turbine system. The efficiency of electrical heater is assumed to be 0.98. The electrical power to be supplied for reaction (3) is :

$$W_3 = \frac{\dot{m}_{CaCO_3}\Delta h_3}{\eta_{heater}} = \frac{45.45\,(\text{mole/s})\times 177.99\,(\text{kJ/mole})}{0.98}$$
$$= 8.26\,\text{MWe}$$

The electrical energy also must be supplied for air blower and slurry pumps. The pumping powers are small. The calculated pumping power of NaOH solution is 0.67 kWe. The calculated pumping power of Na_2CO_3 solution is 0.67 kWe. The calculated pumping power of NaOH-$CaCO_3$ slurry is 1 kWe. The calculated pumping power of $CaCO_3$ slurry is 0.68 kWe. The calculated pumping power of CaO slurry is 0.45 kWe. The calculated pumping power of $Ca(OH)_2$ slurry is 0.64 kWe. The calculated pumping power of H_2O make up is 0.22 kWe. The calculated air blower power is 2.67 kW. Thus the overall air blower pumping power of the capturing CO_2 cycle is 7.00 kWe.

Electrical energy also must be supplied for the operation of mixer agitators in Na_2CO_3 regeneration reactor, $CaCO_3$ regeneration reactor, and CaO-water reactor. The calculated mixer agitator power of Na_2CO_3 reactor is 18.33 kWe. The calculated mixer agitator power of $CaCO_2$ regeneration reactor is 30 kWe. The calculated mixer agitator power of CaO-water reactor is 3.55 MWe. Thus the total mixer agitator power is 46.88 kWe.

Therefore, the total electrical power to be supplied to the atmospheric CO_2 capture system is

$$W_C = W_3 + W_{pupm} + W_{agitator}$$
$$= (8.26 + 0.007 + 0.047)\,\text{MWe} = 8.32\,\text{MWe}$$

3) *Energy balance of methanol synthesis*

The methanol synthesis reaction (reaction (3)) is an exothermic reaction. The energy released is assumed to be unrecoverable. Thus it is no energy needed to supply the reaction (3). However, the reaction (3) needs pressurization until 300 MPa. Thus both of H_2 and CO_3 must be compressed. The electrical energy supply needed for H_2 and CO_2 compression.

The calculated compressor power for CO_2 compression is 0.17 kW. The calculated compressor power for H_2 compression is 5.58 kW. Thus the total electrical power must be supplied for compressor system is

$$W_{compressor} = 5.75\,\text{kWe}$$

4) Overall energy and material balance

The total electrical power needed for overall process is

$$W_{process} = W_E + W_C + W_{compressor} = (110{,}36 + 8.23 + 0.00575)\,\text{MWe} = 118.6\,\text{MWe}$$

Therefore the surplus of electrical power is

$$W_{surplus} = W_{TG} - W_{process} = (205.34 - 118.6)\,\text{MWe} = 86.74\,\text{MWe}$$

The surplus electrical power can be supplied to commercial electric consumers or commercial grid.

This atmospheric CO_2 capture system is assumed to have a capturing capacity of 7.2 ton/h (45.45 moles/s) of CO_2. All of the captured CO_2 is converted to methanol. Thus according to reaction (5), the production capacity of methanol is 45.45 mole/s or 5.24 ton/h of methanol.

To produce 45.45 mole/s of methanol, according to reaction (5), 3 × 45.45 mole/s = 136.35 mole/s oh hydrogen is needed. However, the hydrogen production capacity of the HTSE system powered by 460 MWth PCMSR is 600 moles/s of hydrogen. Thus there is an excess hydrogen production of (600 – 136.35) mole/s = 463.65 mole/s. The 463.65 mole/s or 3.34 ton/h of the excess can be considered as the commercial hydrogen product.

According to reaction (6), the HTSE with a production capacity of 600 moles/s of hydrogen produces 0.5 × 600 moles/s = 300 moles/s of oxygen by-product. The 300 mole/s or 34.56 ton/h of the by-product can also consider as the commercial oxygen product.

4 Summary

It can be summarized that the global warming mitigation is not always an economic burden for human civilization. It is possible to mitigate global warming as well as produce several economics valuable products from this mitigation action. The atmospheric CO_2 capture is the endothermic reaction in overall. Thus the key to this process to be a success is the availability of the abundant, cheap and not CO_2 emitter energy sources. The MSR that utilizes abundant thorium resources is the prominent candidate for the energy source of this system. The analysis has been performed for 460 MWth PCMSR which is coupled with atmospheric CO_2 capture, desalination, hydrogen production and methanol production facilities. For the capturing CO_2 capacity of 7.2 ton/h of atmospheric CO_2, this system can produce valuable outputs of 3.34 ton/h of H_2, 34.56 ton/h of O_2, 5.24 ton/h of methanol and 86.74 MWe of excess electricity.

The results showed that mussel shells contain 14.15% CaO. The optimum yield was 95.12% which was achieved at NaOH concentration of 30% and the catalyst amount of 15%. The biodiesel produced was fulfilled the national standards and had a better quality compared to the conventional homogeneous catalyst. The GC-MS analysis showed that the biodiesel containing methyl ester with the largest component was methyl oleic (61.39%).

References

1. US EPA, Climate Change Science Overview, US Environmental Protection Agency (US EPA), 14 June 2012.
2. W. Cramer, , et al., Executive summary, in Chapter 18: Detection and attribution of observed impacts (archived 18 October 2014), pp.982-984, in IPCC AR5 WG2 A 2014
3. R.N. Stavins (2007). "A U.S. Cap-and-Trade System to Address Global Climate Change" . The Hamilton Project. Retrieved 2014-07-26.
4. C.A. Huff, M.S. Stanford, J. Am. Chem. Soc **133** (45), 18122–18125 (2011)
5. L.C.L.A. Jamshidi, C.M.B.M. Barbosa, L. Nascimento, J.R. Rodbari, Journal Chemical Engineering and Process Technology **4**, 5 (2013)
6. S.T. Hussain,* M. Mazhar, S. Gul, K.T. Chuang,† and A.R. Sanger, B. Kor. Chem. Soc **27** (2006)
7. J.K. Stolaroff, et.al., Environ. Sci. Technol **42**, 2728 – 2735 (2008)
8. G. Collodi, F. Wheeler, Hydrogen Production via Steam Reforming with CO_2 Capture, Milan, Italia, 2009*
9. L. Mingyi, Y. Bo, X. Jingming, C. Jing, J. Power. Sources **177**, 493–499 (2008)
10. J. E. O'Brien, C. M. Stoots, J. S. Herring, M. G. McKellar, E. A. Harvego, M. S. Sohal, K. G. Condie, High Temperature Electrolysis for Hydrogen Production from Nuclear Energy – Technology Summary, The INL is a U.S. Department of Energy National Laboratory, February 2010
11. Nuclear Options for Hydrogen and Hydrogen Based Liquid Fuel Production, MIT Report : MIT-NES-TR-001, September 2003 .
12. A.W. Harto, International Journal of Nuclear Energy Science and Technology **9** (2015)
13. M. Carolina, A. W. Harto, Kusnanto, Pengaruh Variasi Tekanan Operasi Dan Recycling Ratio Terhadap Efisiensi Produksi Hidrogen Pada Molten Salt Reactor (MSR) Dengan Sistem HighTemperature Electrolysis (HTE) Bentuk Planar Cell, Prosiding Seminar Nasional ke-17 Teknologi dan Keselamatan PLTN Serta Fasilitas Nuklir Yogyakarta, 01 Oktober 2011 (in Bahasa Indonesia)

Numerical Study of a Thermosyphon Cooling System: film condensation

Filian Arbiyani[1]

[1]Department of Mechanical Engineering, Universitas Katolik Indonesia Atma Jaya, Jakarta, Indonesia
f.arbiyani@atmajaya.ac.id

Abstract. Studies of condensation in several cooling systems have been conducted. However, the mode of condensation in two-phase cooling systems to achieve a high rate of condensation in compact devices has not been explored. Condensation phenomena, indeed, is a key parameter in designing a thermosyphon water cooled condenser system. The analysis of this condensation phenomena has been done numerically by implementing the governing equations and boundary conditions in commercial MATLAB software. Steady-state laminar film condensation on the radial system is assumed as a condensation phenomenon between vapor and the outer surface of coolant coil. There is a good agreement between experimental and simulation results. Furthermore, for 0.3 LPM 10 °C, it is found the standard deviation of 0.3 %. This small standard deviation indicates the good accuracy of the simulation. At a constant mass flow rate of water, the higher inlet water temperature will result in a higher Nusselt number of water. Furthermore, at the same Nusselt number of water, the lower inlet water temperature obtained a higher film condensation rate. Nusselt number of film condensation increases as the Nusselt number of water decreases at the various constant of mass flow rate of water. Additionally, the lower inlet water temperature will result in a lower Nusselt number of water. The value of Reynold number film condensation increases as Reynold numbers and Nusselt number of water increase. At various constant mass flow rates of the water, at the same Nusselt number of water, the Reynold number of film condensation increases with lower inlet water temperature. The lower inlet water temperature increases the value of Reynold number of film condensation leading to more wavy and turbulent flow. The present study provides guidelines for thermal management engineers to design and fabricate compact cooling systems. **Keywords—film condensation; thermosyphon; numerical study; two-phase cooling system; MATLAB**

1 Introduction

Studies of condensation in several cooling systems have been conducted. However, the mode of condensation in two-phase cooling systems to achieve a high rate of condensation in compact devices has not been explored. Condensation phenomena, indeed, is a key parameter in designing a thermosyphon water cooled condenser system. Since it is difficult to attain a drop-wise condensation, which has higher heat transfer coefficient, and since there was no enhancement structure in the outer condenser wall, thus film condensation was assumed in the fundamental research of the present study.

2 Material and Method

2.1 Governing Equation and Boundary Conditions

Studies of condensation in several cooling systems have been conducted. However, the mode of condensation in two-phase cooling systems to achieve a high rate of condensation in compact devices has not been explored. Condensation phenomena, indeed, is a key parameter in designing a thermosyphon water cooled condenser system. Since it is difficult to attain a drop-wise condensation, which has higher heat transfer coefficient, and since there was no enhancement structure in the outer condenser wall, thus film condensation was assumed in the fundamental research of the present study.

Steady-state laminar film condensation on the radial system (Fig.2) is assumed as a condensation phenomenon between vapor and the outer surface of coolant coil [1]. The governing equations of the pressure drop simulation model of thermosyphon system are the equations of the net driving head caused by the difference between the density of the liquid in the downcomer and that of the vapor/liquid mixture in the riser, which is expressed in equation (1):

$$\frac{dp}{dz} = \frac{dp_f}{dz} + \frac{dp_g}{dz} \quad (1)$$

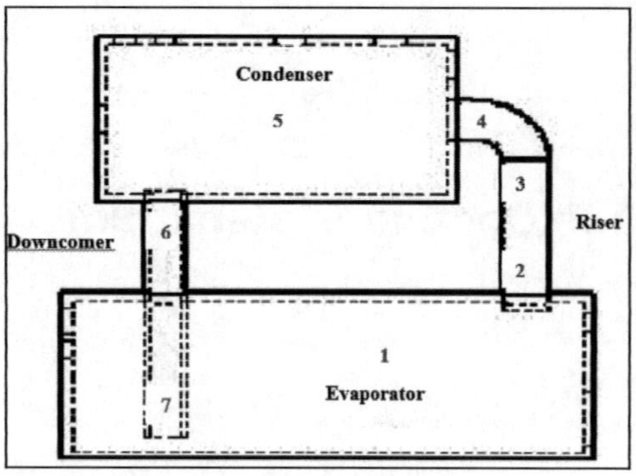

Fig.1. Schematic drawing of thermosyphon system.

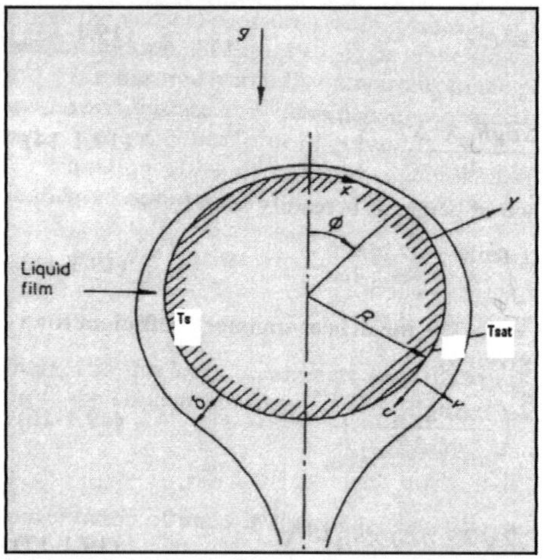

Fig.2. Schematic of film condensation in radial system. [1]

With the boundary conditions:
At $z_i = 0$, $P_7 = P_{two\text{-}phase}$, $P_2 = P_{g,evaporator}$
At $z_i = H$, $P_6 = P_{f,condenser}$, $P_4 = P_{two\text{-}phase}$

The condensation process is assumed as a film condensation in the radial tube (Fig.2), rather than dropwise condensation, as the surface of cooling coil where the condensation occurs is smooth, not rough. Thus the phenomena tend to be a film condensation.

Film condensation phenomena in thermosyphon water cooled condenser system occur at coolant coil surface. When R-113 vapor rises-up to the condenser section, in which water coolant with lower temperature (T_{wi}) than the vapor temperature (T_{sat}) flows over the coolant coil, the temperature of R-113 vapor is reduced below its saturation temperature (T_{sat}). The R-113 vapor then forms to R-113 liquid condensate. This liquid condensate covers the entire coolant coil outer surface, and under the action of gravity, the liquid continuously flows from the surface.

The governing equation for film condensation simulation:

Momentum balance: $\mu_l \dfrac{\partial^2 u}{\partial y^2} + g \sin\theta \Delta\rho = 0$ (2)

Conservation of mass: $\dfrac{\delta}{R}\dfrac{d}{d\theta}\left\{\int_0^{\delta} u\, dy\right\} = \dfrac{k_l \Delta T}{\rho h_{fg}}$ (3)

Energy balance:

$$q = h_{fg}\Gamma = k_l \frac{\Delta T}{\delta},\ \bar{q} = \frac{h_{fg}}{R}\frac{d\Gamma}{d\theta} = k_l \frac{dT}{dy} \quad (4)$$

With boundary conditions:

u = 0 when y = 0; and when y = δ, $\dfrac{\partial u}{\partial y} = 0$

At θ = 0, $\dfrac{d}{d\theta}\left\{\delta^4 \dfrac{\rho \Delta\rho g h_{fg}}{\mu_l k_l d\Delta T}\right\} = 0$

The detail derivations of film condensation characteristics are provided below which can explain the process occurs in thermosyphon system.

Assumptions:

1. The coolant coil surface is maintained at a constant temperature (Ts).
2. The saturation temperature of the quiescent vapor in which the surface is placed is Tsat.

2.2 Fluid Mechanics Consideration

Consider the schematic and fluid element shown.
Assumption:

- Neglect inertia force
- Steady flow
- Viscous force and buoyancy force is balanced

Hence, the equilibrium equation of the fluid element:

$$L_{cc}\delta x \delta y \rho_l g + PL_{cc} - (P + \delta P)L_{cc}\delta y - \tau_{xy}L_{cc}\delta x + \left(\tau_{xy} + \left[\frac{\partial \tau_{xy}}{\partial y}\right]\delta y\right)L_{cc}\delta x = 0 \quad (5)$$

For the quiescent vapor: $\delta P = \delta x \rho_v g$

For a Newtonian fluid: $\tau_{xy} = \mu_l \dfrac{\partial u}{\partial y}$

Thus, by integrating for the velocity, the solution gives the velocity distribution as:

$$u(y) = \frac{g(\rho_l - \rho_v)\sin\theta}{\mu_l}\left[\delta y - \frac{y^2}{2}\right] + C \quad (6)$$

1) Boundary condition

At the coolant coil surface: u(0) = 0, thus C = 0. Hence, the velocity distribution becomes:

$$u(y) = \frac{g(\rho_l - \rho_v)\sin\theta}{\mu_l}\left[\delta y - \frac{y^2}{2}\right] \quad (7)$$

2) Mass flow rate of condensate liquid

The liquid flow rate at a section of the film at a half side is given by:

$$\Gamma\Big|_{\frac{1}{2}side} = \int_0^{\delta} L_{cc}\rho_l u dy \quad (8)$$

Substituting u(y) and then integrate, thus:

$$\Gamma\Big|_{\frac{1}{2}side} = \frac{L_{cc}g(\rho_l - \rho_v)\rho_l \sin\theta}{\mu_l}\left[\frac{\delta^3}{3}\right] \quad (9)$$

$$\text{And thus, } \delta = \frac{L_{cc}^{\frac{1}{3}}\Gamma_{1/2}^{1/3}\mu_l^{\frac{1}{3}}}{\left[\rho_l(\rho_l - \rho_v)g\sin\theta\right]^{\frac{1}{3}}} \quad (10)$$

3) Rate of condensation

The rate of condensation for the half side is defined as:

$$\dot{m}_c = \frac{\partial\Gamma_{\frac{1}{2}}}{\partial x}, \text{ where } x = R\theta \quad (11)$$

4) Film condensation characteristic flow

The characteristic velocity to denote the flow of condensate film is:

$$u_m = \frac{\Gamma}{(\rho_l \delta L_{cc})} \text{ and } D_h = \frac{4A}{p} = \frac{4(\delta \times L_{cc})}{1}, \text{ thus}$$

the Reynolds number is given by

$$\mathrm{Re}_\delta = \frac{u_m D_h \rho_l}{\mu_l} = \frac{4\Gamma}{\mu_l} \quad (12)$$

Therefore, the flow characteristic of film condensation is as follow:

- $\mathrm{Re}_\delta < 30$ is laminar flow
- $30 < \mathrm{Re}_\delta < 180$ is wavy flow
- $\mathrm{Re}_\delta > 1800$ is turbulent flow

2.3 Heat transfer consideration

Assumptions:

- Velocity is small, so that convective energy transfer can be neglected
- Temperature gradient in the θ direction is also small

Energy balance for the small fluid element:

$$qdx = \left[q + \frac{\partial q}{\partial y}dy\right]dx \quad (13)$$

Hence, heat flux at the coolant coil surface is:

$$\dot{Q}_s = -\left(\frac{k_l}{\delta}\right)(T_s - T_{sat})p_2 \quad (14)$$

Heat transfer coefficient is defined as:

$$\dot{Q}_s = \bar{h}_c p_2 (T_s - T_{sat}) \quad (15)$$

From (14) and (15):

$$h_c = \left(\frac{k_l}{\delta}\right) \quad (16)$$

is the expression for the laminar condensation heat transfer coefficient.

The energy balance:

$$\Gamma h_f + h_g \dot{m}_c dx = (\Gamma + d\Gamma)h_f + \dot{Q}_s dx \quad (17)$$

Using relation (11) and from (14):

$$\Gamma_{\frac{1}{2}} = 1.1438\left[\frac{g\rho_l(\rho_l - \rho_v)}{\mu_l^3}\frac{k_l^3(T_{sat} - T_s)^3 R^3}{h_{fg}^3}\right]^{\frac{1}{4}} \quad (18)$$

1) Average heat transfer coefficient and Nusselt number

Hence, for a tube where the condensate flows on both sides and representing the heat transfer rate by the heat transfer coefficient:

$\dot{Q}_s = 2\Gamma_{\frac{1}{2}}h_{fg} = 2\pi R\bar{h}_c(T_{sat} - T_s)$, thus by substituting equation (18), the average film condensation heat transfer coefficient is:

$$\bar{h}_c = 0.728\left[\frac{g\rho_l(\rho_l - \rho_v)}{\mu_l(T_{sat} - T_s)D}\frac{k_l^3 h_{fg}}{}\right]^{\frac{1}{4}} \quad (19)$$

Therefore, the average Nusselt number for a tube of diameter 2R becomes:

$$\overline{Nu} = \frac{\bar{h}_c(2R)}{k_l} = 0.728\left[\frac{g\rho_l(\rho_l - \rho_v)}{\mu_l k_l(T_{sat} - T_s)}\frac{h_{fg}D^3}{}\right]^{\frac{1}{4}} \quad (20)$$

And heat transfer rate:

$$\dot{Q}_s = 0.728x\left[\frac{g\rho_l(\rho_l - \rho_v)}{\mu_l(T_{sat} - T_s)D}\frac{k_l^3 h_{fg}}{}\right]^{\frac{1}{4}} x\, p_2(T_s - T_{sat}) \quad (21)$$

2) Modified latent heat

The work by Rohsenow showed from experiments that the modified latent heat for condensation is given by [2]:

$$h_{fg}^{"} = h_{fg}\left[1 + 0.68Ja\right] \quad (22)$$

Thus, the latent heat which is used in equation (19), (20) and (21) should be as modified latent heat. Hence average

https://doi.org/10.1051/e3sconf/20184201005

heat transfer coefficient, Nusselt number, and heat transfer rate becomes:

$$\bar{h}_c = 0.728\left[\frac{g\rho_l(\rho_l-\rho_v)}{\mu_l(T_{sat}-T_s)D}\frac{k_l^{\,3}h_{fg}^{"}}{}\right]^{1/4} \quad (23)$$

$$\overline{Nu} = \frac{\bar{h}_c(2R)}{k_l}$$

$$= 0.728\left[\frac{g\rho_l(\rho_l-\rho_v)}{\mu_l k_l(T_{sat}-T_s)}\frac{h_{fg}^{"}D^3}{}\right]^{1/4} \quad (24)$$

$$\dot{Q}_s = 0.728x\left[\frac{g\rho_l(\rho_l-\rho_v)}{\mu_l(T_{sat}-T_s)D}\frac{k_l^{\,3}h_{fg}^{'}}{}\right]^{1/4} x\,p_2(T_s-T_{sat}) \quad (25)$$

When using the above expression, all liquid properties are evaluated at the mean temperature between coolant coil surface and the saturated vapor (T_f). Meanwhile, the vapor properties are evaluated at saturation vapor temperature (Tsat).

3) Heat transfer at coolant coil

Heat rate of water flowing in coolant coil is also meant heat dissipation.

$$d\dot{Q}_d = d\dot{Q}_w = \dot{m}_w C_{p,w} dT_w$$

$$\dot{Q}_d = \dot{Q}_w = \dot{m}_w C_{p,w}\int_{T_{wi}}^{T_{wo}} dT$$

$$\dot{Q}_d = \dot{Q}_w = \dot{m}_w C_{p,w}(T_{wo}-T_{wi}) \quad (26)$$

The Nusselt number for water coolant inside the tube is given by:

$$Nu_D = 0.023\mathrm{Re}_D^{0.8}\,\mathrm{Pr}^{0.4},$$

Furthermore, $Nu = \frac{h_w D_i}{k_w}$, thus heat transfer coefficient for water coolant flow inside the tube is:

$$h_w = \frac{k_w(0.023\mathrm{Re}_D^{0.8}\,\mathrm{Pr}^{0.4})}{D_i} \quad (27)$$

4) Overall heat transfer coefficient

Heat transfer coefficient from R-113 vapor to water coolant is determined by overall heat transfer coefficient through the following equation:

$$U = \frac{Di \times h_w \times h_c}{(Do \times \bar{h}_c) + (Di \times h_w)} \quad (28)$$

Furthermore, heat transfer from coolant coil surface to water coolant is determined by:

$$\dot{Q}_w = p_1 h_w(T_s - T_w), \quad (29)$$

Hence, it is necessary to define Tw which is mean temperature of water coolant (Tm). Thus the heat transfer becomes:

$$\dot{Q}_w = p_1 h_w(T_s - T_m) \quad (30)$$

3 Result and Discussion

3.1 Validation

There is a good agreement between experimental and simulation results (Fig.3 and Fig.4).

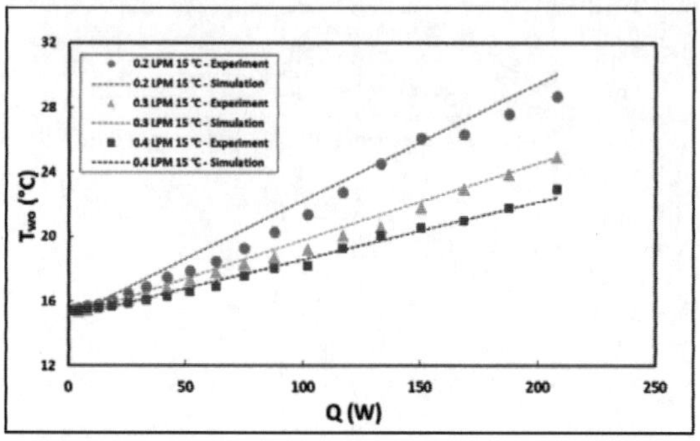

Fig.3. Validation of outlet water temperature at a constant T_{wi} of 10 °C

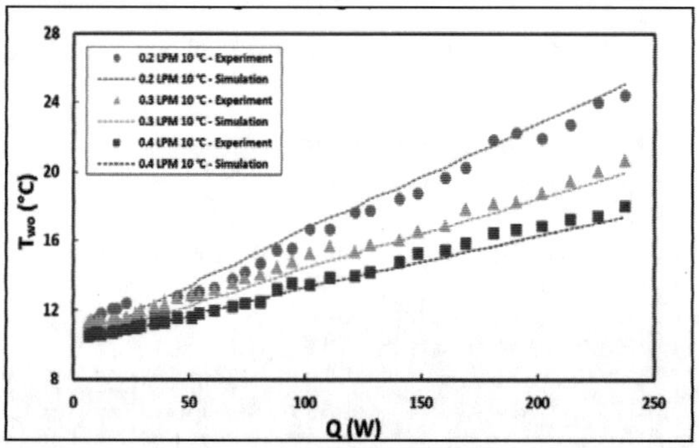

Fig.4. Validation of outlet water temperature at a constant T_{wi} of 15 °C

Furthermore, for 0.3 LPM 10 °C, it is found the standard deviation of 0.3 %. This small standard deviation indicates the good accuracy of the simulation.

3.2 Units Effect of the water mass flow rate (m_w) and inlet water temperature (T_{wi}) on film condensation rate (Γ) Effect of NaOH Concentration to Biodiesel Yield

Effect of the inlet water temperature on film condensation rate (Γ) is presented in Fig. 5, 6 and 7. These figures firstly explained that at constant $\dot{m}_w$, the higher inlet water temperature would result in higher Nusselt number of water. Furthermore, at the same Nusselt number of water, the lower inlet water temperature obtained a higher film condensation rate.

https://doi.org/10.1051/e3sconf/20184201005

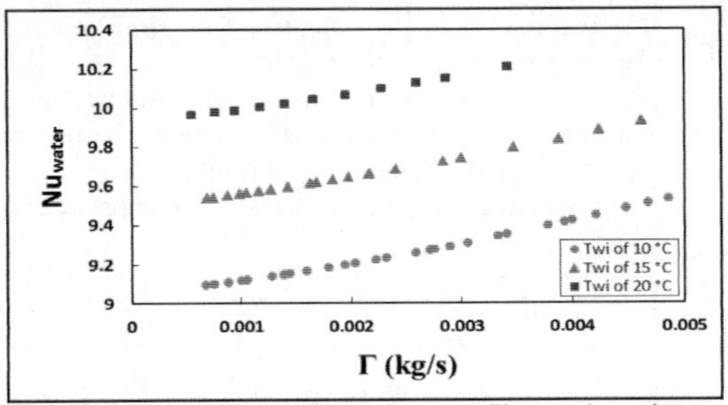

Fig.5. Effect of inlet water temperature on film condensation rate at constant alidation of outlet water temperature at a constant m_w of 0.0033 kg/s.

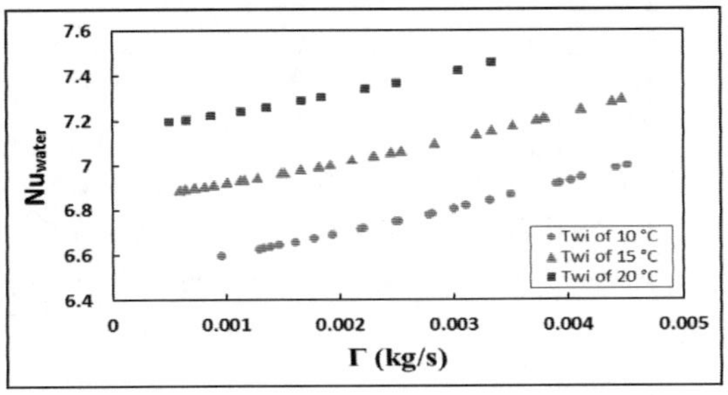

Fig.6. Effect of inlet water temperature on film condensation rate at constant alidation of outlet water temperature at a constant m_w of 0.005 kg/s.

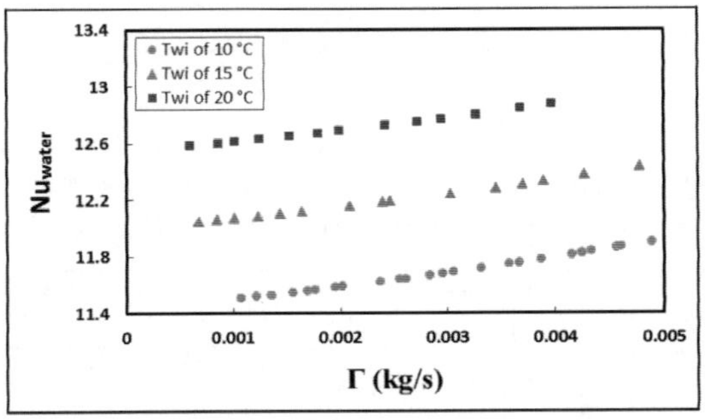

Fig.7. Effect of inlet water temperature on film condensation rate at constant alidation of outlet water temperature at a constant m_w of 0.0067 kg/s.

Figures 8, 9, and 10 present the effect of the mass flow rate of inlet water on film condensation rate. It can be noticed that Reynold number of water increased as the mass flow rate of inlet water increased. Moreover, it also can be noticed that the larger Reynold number will increase the film condensation rate.

In the present study, it can be concluded that the Nusselt number of water is 6.5-12.87, Reynold number of water is 389 to 1069.3, and film condensation rate is 0.0005 kg/s to 0.0048 kg/s.

This study assumed that the condensation process formed only as film condensation and dropwise condensation was not considered. This is due to the fact that the condenser tube surface is smooth, which tends to favour film condensation. Nevertheless, the results are significant to provide condensation characteristics of thermosyphon cooling systems.

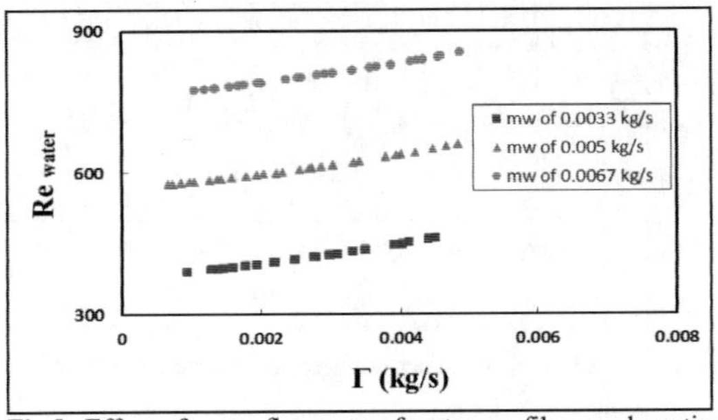

Fig.8. Effect of mass flow rate of water on film condensation rate at a constant T_{wi} of 10 °C.

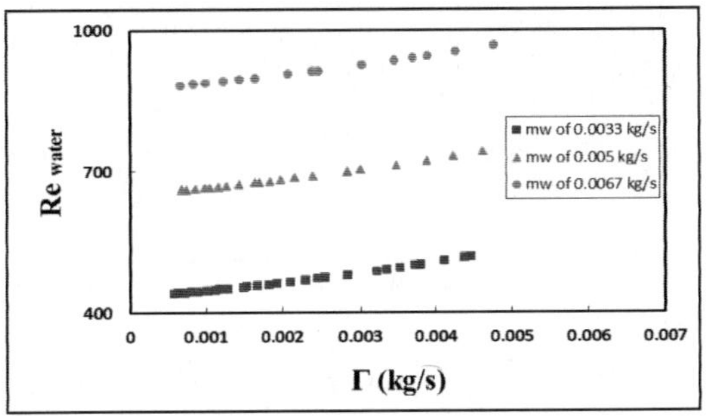

Fig.9. Effect of mass flow rate of water on film condensation rate at a constant T_{wi} of 15 °C.

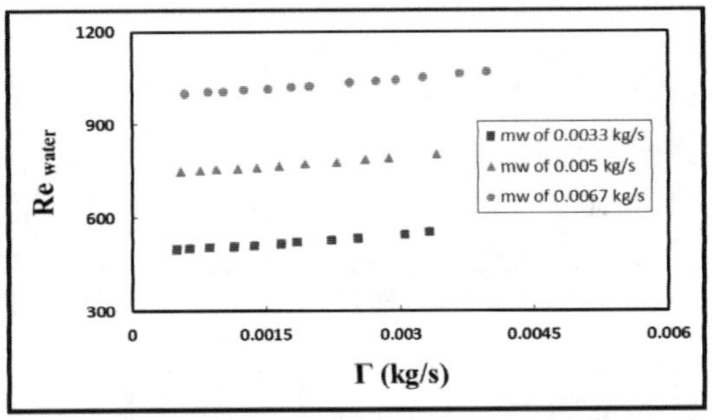

Fig.10. Effect of mass flow rate of water on film condensation rate at a constant T_{wi} of 20 °C.

3.3 Effect of the Water Mass Flow Rate (m_w) and Inlet Water Temperature (T_{wi}) on Nusselt Number Film Condensation (*Nu*)

Figures 11, 12, and 13 show that Nusselt number of film condensation increases as the Nusselt number of water decreases at the various constant of m_w . Additionally, the lower temperature T_{wi} will result in lower Nusselt number of water. It may be observed that higher Nusselt number of film condensation is obtained with by flow of lower T_{wi}.

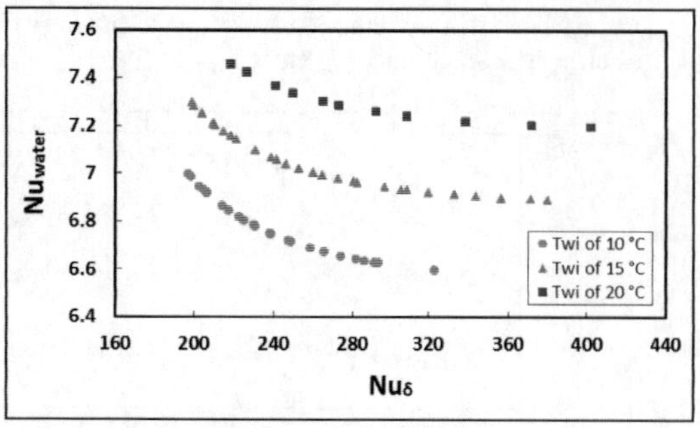

Fig.11. Effect of inlet water temperature on the Nusselt number of film condensation at a constant $\dot{m}_w$ of 0.0033 kg/s.

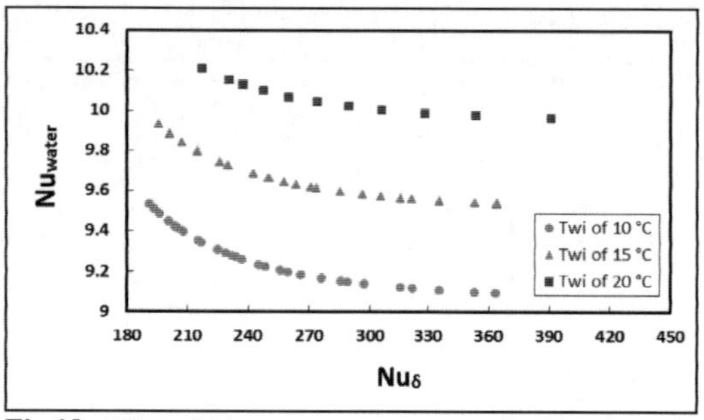

Fig.12. Effect of inlet water temperature on the Nusselt number of film condensation at a constant $\dot{m}_w$ of 0.005 kg/s.

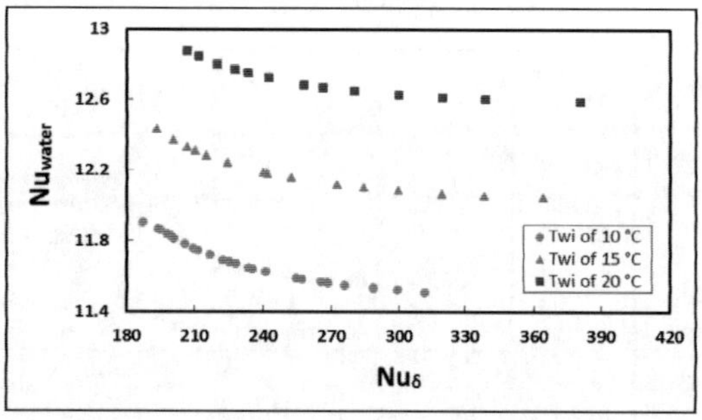

Fig.13. Effect of inlet water temperature on the Nusselt number of film condensation at a constant $\dot{m}_w$ of 0.0067 kg/s.

The effect of the water mass flow rate ($\dot{m}_w$) on Nusselt number film condensation (Nu) at various constant T_{wi} can be shown in Fig. 14, 15 and 16. It can be noticed that the higher Nusselt number film condensation can be gained by lower Reynold number of water. In addition, the lower Reynold number of water can be gained by flowing lower mass low rate of inlet water. Hence, the lower mass low rate of inlet water will result in higher Nusselt number film condensation. This result is in conflict with the requirement of achieving high cooling capacity and large film condensation rate. It explains that an increase in the film condensation rate, which represents an increase in the thickness of the film, causes a slight reduction of the average convection coefficient. This slight reduction in the average convection coefficient, therefore, obtains a lower Nusselt number of film condensation. These results, which are in dimensionless number forms, provide significant data which are applicable for any geometry of cooling systems. However, this study was limited to water as an external coolant.

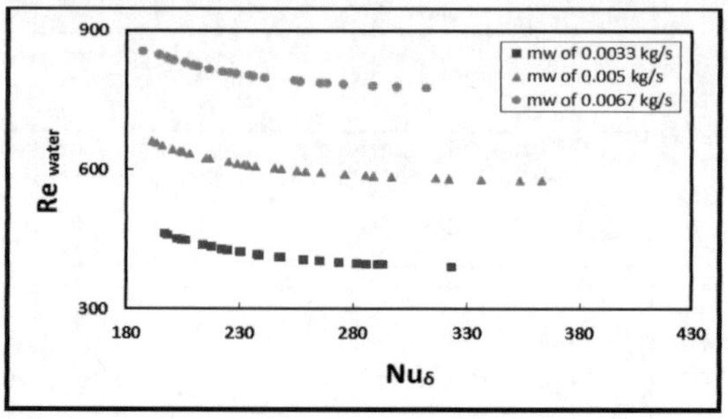

Fig.14. Effect of mass flow rate of inlet water on the Nusselt number of film condensation at a constant T_{wi} of 10 °C.

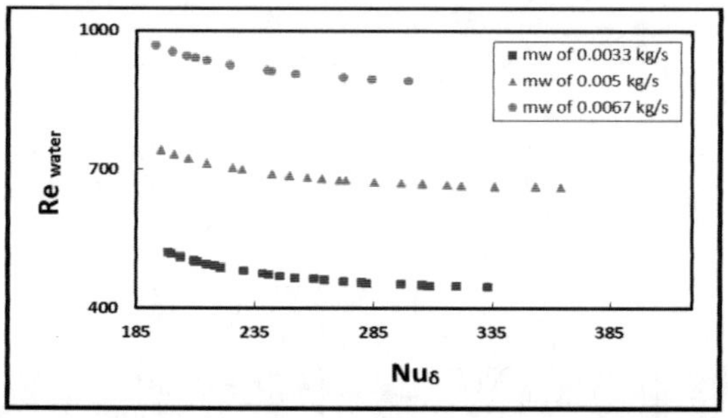

Fig.15. Effect of mass flow rate of inlet water on the Nusselt number of film condensation at a constant T_{wi} of 15 °C.

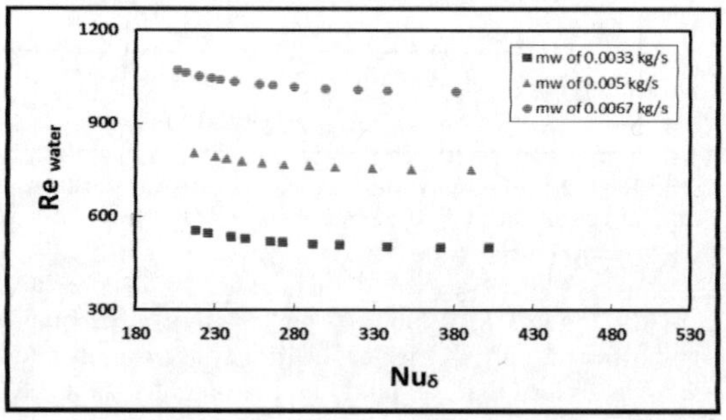

Fig.16. Effect of mass flow rate of inlet water on the Nusselt number of film condensation at a constant T_{wi} of 20 °C.

3.4 Effect of the Water Mass Flow Rate ($\dot{m}_w$) and Inlet Water Temperature (T_{wi}) on Reynold Number Film Condensation (Re_δ)

The flow regime of film condensation is determined by the Reynold number. For laminar flow, $Re_\delta < 30$; for wavy-transition, $30 \le Re_\delta \le 1800$; and for turbulent flow, $Re_\delta > 1800$ [3].

The film condensation layer in the present study is in the laminar and wavy-transition regime. It can be seen from Fig. 17-22 that the Re is $3.9 \leq Re_\delta \leq 29$ for the laminar regime and $30 \leq Re_\delta \leq 40$ for the wavy-transition regime. It is found by flowing inlet water temperature of 20 °C with the mass flow rate of 0.0033 kg/ and 0.005 kg/s will result in laminar regime. Inlet water temperature of 20 °C and mass flow rate of 0.0067 kg/s has a laminar regime in the range heat flux of 0.37 W/cm^2 – 2.97 W/cm^2. The detailed heat flux range (W/cm^2) for laminar and wavy-transition regime can be seen in Table 1:

Table 1. Heat Flux Range For Laminar and Wavy-Transition Regime

$\dot{m}_w$ (kg/s)	T_{wi} (°C)	Laminar (W/cm^2)	Wavy-transition (W/cm^2)
0.0033	10	0.7 – 3.37	3.62 – 4.74
	15	0.37 – 3.2	3.37 – 4.74
	20	All range	-
0.005	10	0.46 – 3.2	3.37 – 4.74
	15	0.37-3.37	3.62 – 4.74
	20	All range	-
0.0067	10	0.7 – 3.2	3.37 – 4.74
	15	0.37 – 2.97	3.2 – 4.74
	20	0.37 – 2.97	3.2 – 4.74

It also can be noticed that the value of Reynold number film condensation increases as the Nusselt number of water increase as can be seen in Fig. 17, 18, and 19. It also can be seen that at various constant mass flow rates of the water, at the same Nusselt number of water, the Reynold number of film condensation increases with lower inlet water temperature. Therefore, it can be observed that lower inlet water temperature increases the value of Reynold number of film condensation leading to more wavy and turbulent flow.

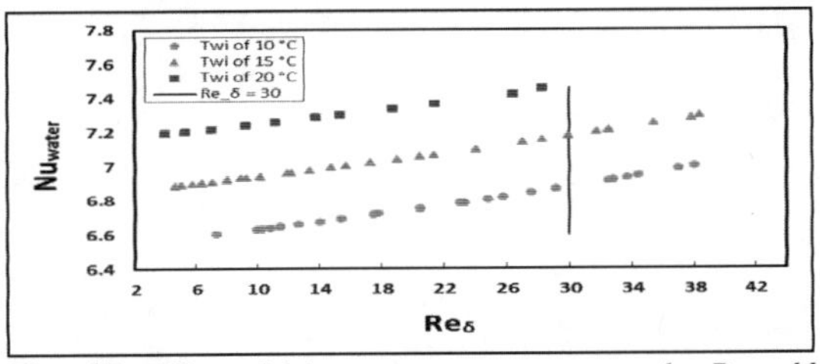

Fig. 17. Effect of inlet water temperature on the Reynold number of film condensation at a constant $\dot{m}_w$ of 0.0033 kg/s.

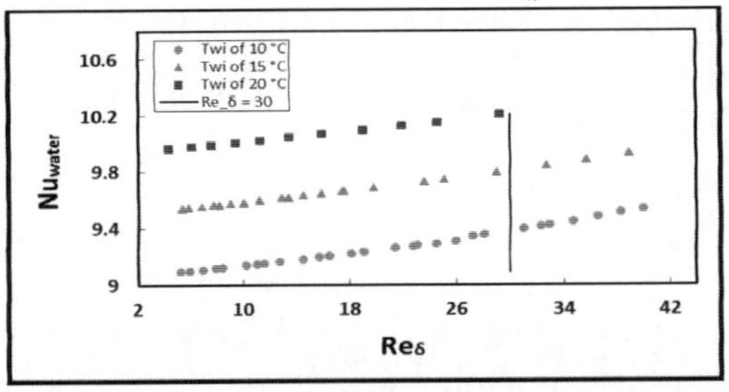

Fig. 18. Effect of inlet water temperature on the Reynold number of film condensation at a constant $\dot{m}_w$ of 0.005 kg/s.

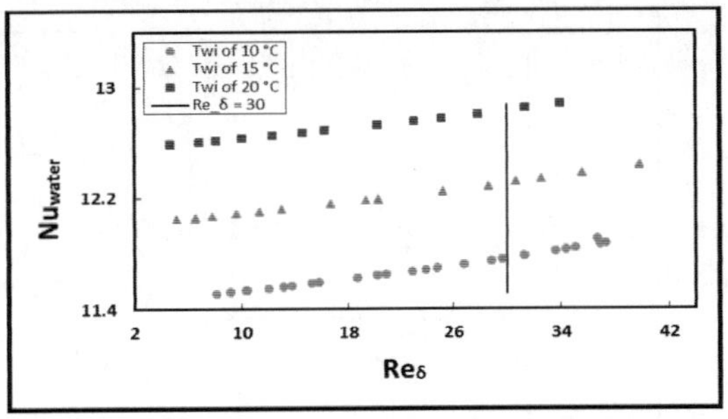

Fig. 19. Effect of inlet water temperature on the Reynold number of film condensation at a constant $\dot{m}_w$ of 0.0067 kg/s.

Figures 20, 21, and 22 show that the values of Reynold number of film condensation increase as the Reynold numbers of water increase. In addition, the higher mass flow rate of water will result in higher Reynold number of water. Therefore, it also can be noticed that greater $\dot{m}_w$ increases the value of Re which also leads to more wavy and turbulent flow.

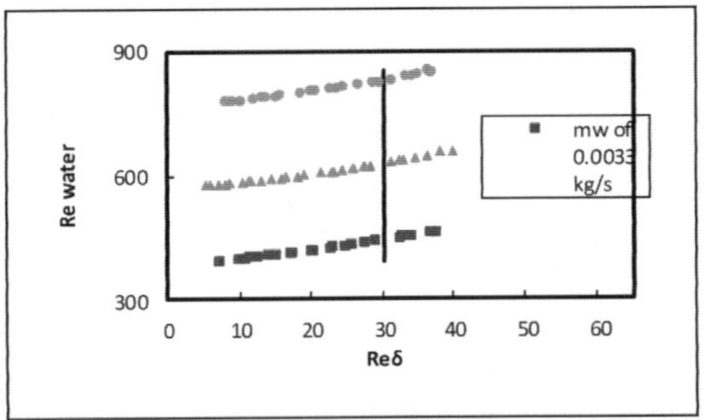

Fig. 20. Effect of mass flow rate of inlet water on Reynold number of film condensation at constant T_{wi} of 10 °C.

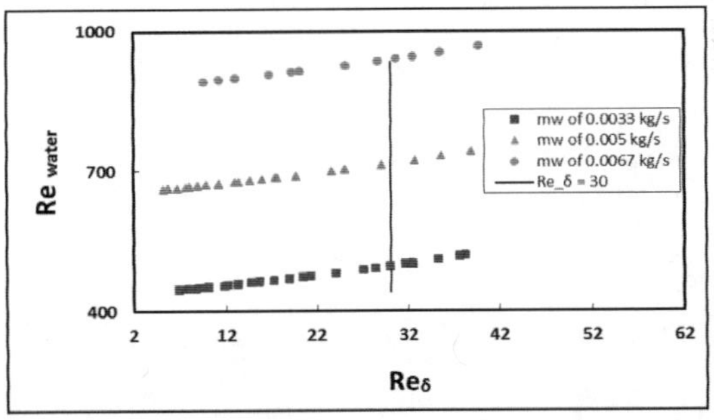

Fig. 21. Effect of mass flow rate of inlet water on the Reynold number of film condensation at a constant T_{wi} of 15 °C.

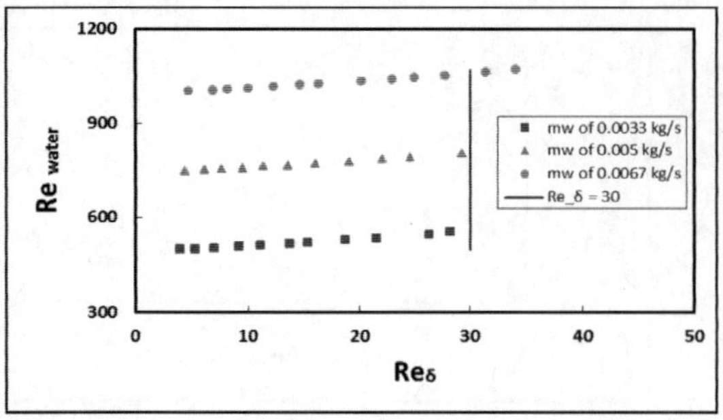

Fig. 22. Effect of mass flow rate of inlet water on the Reynold number of film condensation at a constant T_{wi} of 20 °C.

4 Conclusion

The study was conducted both numerical and experimental investigation. There is a reasonable agreement between numerical predictions and experimental data owing to the small standard deviation of 0.3 %.

The results of the present study, which is in dimensionless number forms (Reynold and Nusselt number), provide significant data which are applicable for any geometry of cooling systems. Even though this study was limited to water only as an external coolant, the present study can be used as a guideline for thermal management engineers to design and fabricate a compact cooling system.

References

[1] F. Arbiyani, ARPN Journal of Engineering and Applied Sciences **11**, 4016-4021, (2016)

[2] F.P. Incropera, D.P. Dewitt, T.L. Bergman, A.S. Lavine, Fundamentals of Heat and Mass Transfer Sixth edition, 645-652 (John Wiley and Sons (Asia), Singapore, 2007)

[3] S.G. Kandlikar, M. Shoji, V.K. Dhir, *Handbook of Phase Change: Boiling and Condensation*, (Taylor&Francis, Philadelphia, 1999)

The Effect of Temperature Variation on Water Scrubber System to Optimize Biogas Purification

Arief Abdurrakhman[1], *Dhirga* Kurniawan[2], and *Moch* Machrus Adhim[2]

Department of Engineering Physics, Institut Teknologi Sepuluh Nopember, Surabaya, Indonesia
[1]ariefabdurrakhman@gmail.com, [2]dhirgak@gmail.com, [3]adhimmachrus@gmail.com

Abstract. Biogas is made from the anaerobic process, which methane (CH4) as a primary element with some other elements. The composition of biogas in the experiment still contains H2S gas pollutants by 41,76 ppm. The amount of H2S content can reduce CH4 composition, so it has a lot of efforts to reduce H2S in biogas. One of them used water scrubber system, but the water scrubber system always changed according to the temperature ambient. It can cause inconsistency biogas quality, which is the reduction of H2S always changed. Another study showed that the speed of the water flow of 5.5 to 11 liters / min and water temperature of 10-25 °C, can reduce the H2S content in biogas up to 4.5 - 8%, of the initial content by 37,8% to 43.6%. So that is needed the design of water cooling system on purifying biogas. While this research aims that the lower water temperature will increase the effectiveness of the reduction of H2S with water temperature 15-20 °C. From the experiments that have been carried out test results obtained when the water temperature 20 °C the content of H2S was 0.7 ppm when the water temperature 17°C H2S content was 0.6 ppm when the water temperature 15 ° C H2S was 0.5 ppm. **Keywords— absorption; biogas; H2S; purification; water cooling system**

1. Introduction

Gas purification greatly affects the quality of the biogas produced. The quality of biogas depends on the concentration of methane gas (CH_4) are contained. The higher of methane gas content in biogas, make the higher the quality of biogas, and vice versa. In addition, the quality of biogas can be improved by performing the purification of biogas, aim to reduce the levels of hydrogen sulfide (H_2S) and carbon dioxide (CO_2) [1]. Levels of H_2S contained in the biogas can be harmful to users because at high levels or more than 500 ppm can cause lung damage and even lead to death [2]. So that is necessary purification process of biogas for an energy sustainability quality alternative, non-hazardous, and environmentally friendly.

In addition, the utilization of biogas for electrical purposes is still not used optimally. In a direct interview with Mr. Hariyanto (Consultant Biogas in Nongkojajar, Pasuruan Sub-District, East Java Indonesia), he said that the lack of biogas utilization as a source of electricity supply is caused by H_2S gas content are still relatively high, ranging from 40-45 ppm, while the minimum tolerance limit concentration of H_2S is 5 ppm [3]. So that, a gas purification system is needed to produce more pure of biogas, i.e., with a high content of CH_4, and H_2S content is low.

Currently, there are several biogas purification methods have been developed. There are absorption, adsorption, cryogenic, membrane, and carbon mineralization technology [4]. Literally, gas-liquid absorption method with water scrubber system is a priority in biogas purification applications, because its main ingredient is water can be relatively cheap and available, and environmentally friendly [5]. In water temperature of 10-20 °C, CO_2 gas solubility level between 2.5 to 1.6 g per kg of liquid gas. While at the same temperature value will be able to dissolve the H_2S gas at 5.5 to 3.9 g per kg of liquid gas [6]. One of the most widely used and economical is the incorporation of adsorption and absorption method, which biogas purification work using a water spray particles to bind water to gas impurities, including H_2S gas [1]. Biogas purification from acid gasses CO_2 and H_2S using water scrubber reduced more H_2S than CO_2 [7].

In this research will be optimized absorption of H_2S gas to the temperature in the biogas purification system with a water scrubber. It is known that the temperature ambient of water scrubber always changing, through mathematical modeling approaches will obtain the absorption of H_2S gas at the temperatures between 20-15°C of water scrubber plant in the area Nongkojajar Pasuruan Sub-District, East Java Indonesia.

https://doi.org/10.1051/e3sconf/20184201006

2. Method

2.1 Water Scrubber

Water Scrubber uses liquid fluid in the working process. Most of the fluid is water. This is the difference between Dry Scrubber with Wet Scrubber. The use of these fluids gives advantages to the system because it can control Scrubber Water-soluble gasses such as SOx, NOx, H2S, which in Dry Scrubber these pollutants cannot be controlled properly. Other advantages are due to the temperature control system Scrubber Water merged two different types of fluid temperature. Water for ways of working more efficiently than Dry Scrubber because it is able to capture particles as small and can bind gasses. The principle is to drain the liquid fluid in the flow of gas so instantly filtered by the liquid fluid [6].

2.2 Reaction Equilibrium

According to [4] absorption process of H_2S gas can be done with water (H_2O). This method is considered very economical because it only uses water as raw material, which availability is very abundant. To do H_2S gas absorption optimization research needs to be done mathematical modeling. The aim of this mathematical modeling is to determine the objective function of the variables to be optimized.

The process of absorption of H_2S in biogas chemical reactions described in the following equation:

$$H_2S_{(g)} + H_2O_{(l)} \leftrightarrow H_2S_{(l)} \quad (1)$$

$$H^+ + HS^- \leftrightarrow H_2S_{(l)} \quad (2)$$

$$H^+ + S^{2-} \leftrightarrow HS \quad (3)$$

According to the aspect of kinetics reaction, H_2S gas absorption by the absorbent solution can be categorized as a heterogeneous reaction that is gaseous and liquid [8]. Based on the equation 3, the obtained mass balance is accompanied by a chemical reaction as follows:

$$A + B \leftrightarrow C \quad (4)$$

2.3 Mass Balance

According to William Nock et al., 2011, the mass transfer of data with the reaction can be determined using mass transfer scheme as Fig. 1 [9].

Based on the schematic of Fig. 1 it was mathematically obtained mass balance in gas, stating that the mass balance is accumulated in the gas or liquid phase to follow the following equation [9]:

$$Accumulation = input\text{-}output - reaction \quad (5)$$

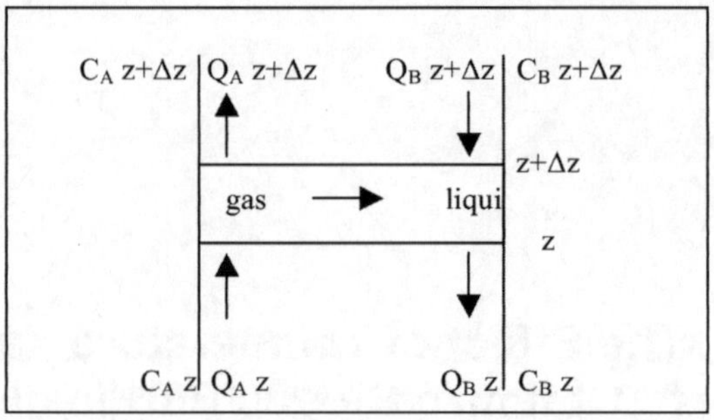

Fig. 1. Mass Transfer Scheme on Absorption Process.

2.4 Solubility of H_2S

Reduction of H_2S officially has been made by some researchers. Fig. 2 shows the solubility model to the temperature of water [6]:

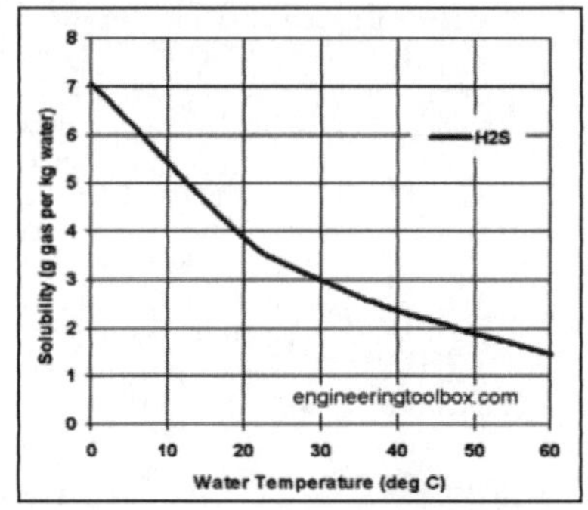

Fig. 2. The solubility of H_2S to Water Temperature.

2.5 Design of Biogas Purification System

This study proposed that the usage of water scrubber system with 4 temperature variations. H_2S content in biogas is measured at the input and output column purification. Biogas is used in the outcome of an anaerobic digester with base material cow dung. Anaerobic digester reactor types used concrete dome reactor (fixed-dome). Biogas reactor size used volume 8 m^3. Plant of experimental study shown in Fig. 3.

2.6 Efficiency of H_2S Absorption

Through the H_2S gas concentration value, input and output are already known, the efficiency of absorption of H_2S gas can be known through calculation% recovery by the following equation, [9]:

$$\%\ recovery = H_2S\ output : H_2S\ input \quad (6)$$

Percent recovery represents the amount of H_2S gas particles are absorbed by water scrubber. The higher of recovery value, make the better efficiency, vice versa.

https://doi.org/10.1051/e3sconf/20184201006

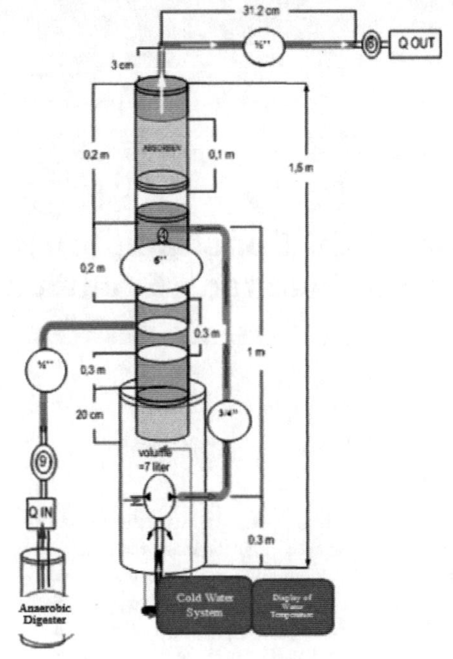

Fig. 3. The design of Biogas Purification System.

3. Results and Discussion

3.1 Water Scrubber Temperature Effect to Reduce H_2S in Biogas

Tool cooling design can hold as much water as 1,300L, and can cool the water in accordance with a predetermined set point of 15 ° C, to cool the water in order to reach the setpoint can be found under the monitoring results are stored in a database of stored data required term time during 18 minutes. The main function of the cold water is to reduce gas impurities contained in biogas, especially H_2S. In the cold water, the experiment has 3 different temperature variations. Here are the data obtained.

The mechanism of water temperature measurement using thermocouples connected to the microcontroller to run the pump and solenoid valve actuators automatically. When the cooling device operation, it will Peltier cool box filled with water for the purification process water scrubber system.

Table 1. The Effect Of Temperature To Reduce H_2S

No	Temperature (^{0}C)	H_2S Measurement Result (ppm)	Method
1	20	0.7	Ionization Merk Photovac Type 2020 ppb pro
2	17	0.6	
3	15	0.5	

The result of this research is a reduction of the optimum value of the H_2S which the water temperature in a water scrubber at 15 °C, amounted to 98.7%. Through the reduction of pollutant gas levels in this study, it could theoretically increase the level of methane in the biogas and methane content of the largest set of conditions in a water scrubber water temperature of 15 °C. Results of testing the duration of the cooling process to achieve each set point are as follows: T = 15 °C for 60.6 seconds; T = 17 °C for 61.1 seconds, and T = 18 °C for 62.1 seconds.

4. Conclusion

The proposed water scrubber operation with cooling water was shown to be successful in the reduction of H_2S in biogas. The lower the temperature of the water in the water scrubber system, also increasing the reduction of levels of H_2S.Through this H_2S reduction, then theoretically methane levels will increase and improve the quality of biogas.

Thanks to the family of authors, advisors, lectures, and all friends who supported this research. Without you maybe it was not successful.

Notations

A = Surface absorber area, (m^2)
C_A = H_2S gas concentration in liquid, (ppm)
C_{Ai} = H_2S gas concentration in interface (ppm)
C_B = Concentration of H, (ppm)
C_C = Concentration of S, (ppm)
D = Absorber diameter,(m)
G = Biogas flow rate, (liter/minute)

References

1. A. Abdurrakhman, M. M. Adhim, T. Ramadhan, ICAMIMIA, 86-90 (2015)
2. A. Noyola, J.M. Morgan-Sagastume, J.E. López-Hernández, Rev. Environ. Sci. Bio **5**, 93-114 (2006)
3. A. Wellinger, A. Lindeberg, "Biogas upgrading and utilization," IEA Bioenergy, Task 24 : Energy from biological conversion of organic waste, Nova Energie GmbH.
4. M. O'Brien, Food Industry Environment Conference (1991)
5. W.M. Budzianowski, C.E. Wylock, P.A. Marciniak, Energ. Convers. Manage (2016)
6. L. Medard, Gas Encyclopedia (1976)
7. M. Islamiyah, T. Soehartanto, R. Hantoro, A. Abdurrahman, KnE Energy, **2** (2), 126- 131 (2015)
8. F. Osorio, J.C. Torres, Renew. Energ., 2164-2171 (2009)
9. W. Nock, S. Heaven, C. Banks, "Modeling water scrubbing biogas based technology," Promotion of Biogas Upgrading and Bottling in India and Europe Union, University of Southampton, 2013

Fuel Loading Pattern Optimization with Constraint on Fuel Assembly Inventory Using Quantum-Inspired Evolutionary Algorithm

M. Rizki Oktavian[1], *Alexander* Agung[1], *Andang* Widi Harto[1]

[1]Department of Nuclear Engineering and Engineering Physics, Universitas Gadjah Mada, Yogyakarta, Indonesia

Abstract. Nuclear fuel management was done by optimizing fuel loading pattern in a reactor core. Practically, performing fuel loading pattern optimization was difficult because of its combinatorial problem complexity which needed to be solved. Therefore, Quantum-inspired Evolutionary Algorithm (QEA) which could solve the combinatorial problem faster than conventional method was used. The main purpose of this research was to obtain an optimum fuel loading pattern of KSNP-1000 reactor core without altering fuel assembly inventories. KSNP-1000 core was modeled in SRAC code package using PIJ module for fuel pins and fuel assemblies' lattices and CITATION module for fuel assemblies' pattern in a quarter core symmetry. Optimization problem adaptation using QEA was made by presenting 52 fuel assemblies in Q-bit individuals with the length of 8 Q-bits. Q-bits were converted to corresponding bit values and then given weight which would be used as consideration to optimize the pattern. The optimization program was coupled with the SRAC neutronic code to obtain the values of effective multiplication factor (k_{eff}) and power peaking factor (PPF). The optimization was calculated based on fitness value which was a function of k_{eff} and PPF values with the particular weight factor. Using a rotation gate angle of $\Delta\theta$=0.02π and a weight factor of w=0,041, fuel loading pattern optimization was done on 360 days burnup level. The optimization resulted in k_{eff} and PPF value of 1.11233 and 1.944 respectively. By calculating k_{eff} value on various burnup levels for the chosen core loading pattern, reactor cycle length obtained was 659 days with PPF at BOC was 2.19. Compared to the standard KSNP-1000 core which had 560 days of cycle length, the optimized core configuration increased 17.67% in cycle length. **Keyword—Optimization; fuel loading pattern; QEA; cycle length; keff; PPF**

1. Introduction

Nuclear fuel management includes ways and considerations to decide things to do in order to control flux and reactivity of a reactor core and spatial power distribution and burn-up of nuclear fuels. It has purpose primarily to minimize operation cost of a nuclear power plant but fulfilling the operation requirement [1].

Generally, nuclear fuel management can be divided into two parts, out-of-core and in-core. Out-of-core fuel management focuses on questions "What to buy?" and "What to be reinserted to the core?" and multi-cycles planning [2]. On the contrary, in-core fuel management has an objective to decide loading pattern in the core, e.g., fresh fuels, spent fuels and burnable poisons and control rods location [1].

In-core fuel management is a problem need to be kept attention in the uses of nuclear energy. Practically, it is done by optimizing fuel loading pattern in a reactor core. An optimum condition in the reactor core can be achieved by looking at its cycle length and power distribution.

Optimization in the nuclear reactor core can be quantified by its effective multiplication factor (k_{eff}) for certain burnup step. Besides, safety factor must also be considered since optimization may affect the power distribution. For safety reason, PPF or power peaking factor must be used as a constraint in optimizing reactor core. The power peaking factor in the nuclear reactor is defined as the ratio of maximum power to average power in the reactor core. The best PPF is unity, means that all power generation perfectly distributed in the core with exact value.

Practically, fuel loading pattern optimization is difficult to be done because of the excessive numbers of fuel assemblies in a nuclear reactor. In a standard PWR, at least there is more than 10^{25} combination to try. The combinatorial problem then will be complicated and need a bunch of time if it is done using conventional algorithm [3].

Evolutionary Algorithm (EA) is one kind of algorithm used to solve loading pattern (LP) optimization in a nuclear reactor. Basically, EA is an optimization method and stochastic searching tool based on natural biological evolution. Compared to that conventional method, e.g., calculus-based and enumerative strategic, EA is more robust, global and applicable to most of the cases [4], [5].

Along with the EA, quantum computing also shows a great attention in the world as a modern computation method. Quantum computing is a result of an interdisciplinary study of computer science and quantum physics [6]. Because of its strength and fast working ability, a lot of researches and developments happened in quantum mechanics computer since its introduction on the 1980s [5 , [7].

Compared to Genetic Algorithm (GA), a former evolutionary algorithm, Quantum-Inspired Evolutionary Algorithm (QEA) produces better performance with a lesser amount of calculations [3]. However, QEA is not a quantum algorithm, but it is a shape of EA adapted with quantum computing on a digital computer. Adaptation of quantum computing in metaheuristic method of QEA is predicted to afford faster computation and more accurate solution of the combinatorial problem [5]. The explanation of QEA origin can be illustrated in Fig.1.

Fig. 1. QEA illustration on the digital computer [4].

A comparison of GA and QEA methods was conducted by performing both algorithms on Knapsack problems [6]. The experiment resulted that on 100 items, both algorithms showed the same value for every problem. However, on 500 items case, QEA performed better result than GA [6].

QEA method was also successfully performed to optimize Brazilian Pressurized Water Reactor (PWR) nuclear power plant, Angra 1 [1]. The calculation indicated that QEA produced a better result with a lesser amount of evaluations (70,400 evaluation) than classical metaheuristic optimization methods. Moreover, compared with another quantum-inspired method, Quantum Population-Based Incremental Learning (QPBIL), QEA resulted a higher average of boron concentration.

Quantum computing uses the smallest unit to store information, named *quantum bit* or Q-bit [4]. A Q-bit can be in a state of "1" or "0" or even superposition of both. The State of Q-bit can be described as :

$$|\Psi\rangle = \alpha|0\rangle + \beta|1\rangle \quad (1)$$

where α and β are complex numbers indicating a probability of a related state. $|\alpha|^2$, on one hand, expresses a probability that Q-bit will be found in the state of "0" and $|\beta|^2$, on the other hand, the express probability of Q-bit to be found in state of "1" [4]. Normalization of both in a unity condition produces:

$$|\alpha|^2 + |\beta|^2 = 1 \quad (2)$$

Q-bit can be altered by an operator named *quantum gate* or Q-gate [4]. Q-gate is a reversible computation operator to make Q-bit update process fulfilling normalization condition as stated on Equation (2). Q-gate used in QEA is stated as follow:

$$U(\Delta\theta_i)\begin{bmatrix}\cos(\Delta\theta_i) & -\sin(\Delta\theta_i)\\ \sin(\Delta\theta_i) & \cos(\Delta\theta_i)\end{bmatrix} \quad (3)$$

with θ_i, i= 1, 2, 3, …, is a rotation angle of each Q-bit to "0" or "1" state based on its sign. Rotation angle increment, $\Delta\theta_i$, must be designed to adapt the case. Based on some works, the value of $\Delta\theta_i$ is recommended in the range of 0,001 to 0,05 [5].

When performing optimization, QEA updates the objective function value each time the higher value found and migrate it to the result. When the migration conducts between individual (a series of Q-bits) in a population (a group of individuals), it is called *local migration.* When it happens to the population of the same generation (a time step of repetition), it is named *global migration*.

Basically, the optimization of k_{eff} and *PPF* is a multiobjective problem. We want k_{eff} as high as possible while PPF as low as possible, within constraints. However, to simplify the calculation, the multiobjective optimization can be transformed into a single objective optimization problem. This can be done by introducing a *fitness* variable which connects the parameters involved. In our optimization problem, the fitness function uses *PPF* as penalty component to compromise the value of k_{eff}. Value of *fitness* as the objective function is stated as follow:

$$fitness = k_{eff} - w \times \Delta PPF \quad (4)$$

with

$$\Delta PPF = PPF - PPF_{standard} \quad (5)$$

where w is the function's weight factor. The value of $PPF_{standard}$ is set to 2, obtained from technical data of KSNP-1000 as reported in the BATAN/KEPCO document in 1997 [9].

Korean Standard Nuclear Power Plant (KSNP) is PWR-type nuclear reactor which is designed based on Gen II technology [8]. KSNP-1000 was used to be a collaboration study of National Nuclear Energy Agency (BATAN) and Korean Electric Power Company (KEPCO) on 1997 [9], [10]. Subcomponents of KSNP-1000 are illustrated in Fig. 2.

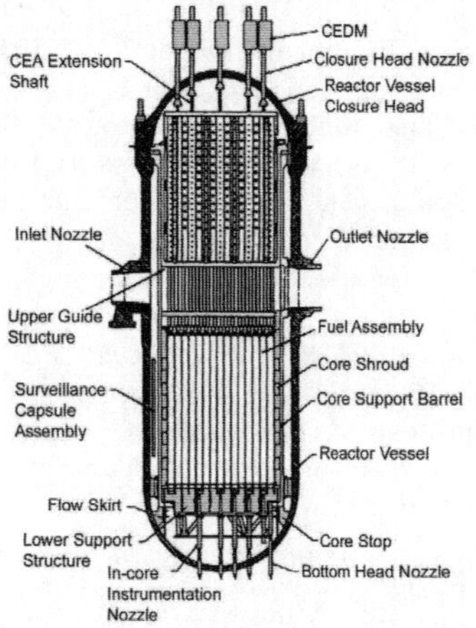

Fig. 2. Subcomponents of KSNP-1000.

Neutronic computation on KSNP-1000 fuels was performed using SRAC95 code system [8]. The computation resulted in the conclusion that multiplication factor of KSNP-1000 core is affected by the amount of U235 enrichment proportionally. Based on the same work, the best enrichment of KSNP-1000 fuel is 5% resulted in 47,000 MW-day/ton of burn-up and 1.01140 of k_{eff} [8].

KSNP-100 optimization had been conducted using GA method with only three types of fuel rods. They were enriched with 1.5%, 2.5% and 3% of ^{235}U. The numbers of each type of fuel can be altered in each optimization step [9,10]. However, the model used in work was not the based on the standard core of KSNP-1000 in types and numbers of fuel assemblies. Therefore, to get the accurate result of KSNP-1000 optimization, a model based on actual reactor core must be used. Moreover, fixed fuel assemblies have to be considered as well.

Based on those considerations, using QEA method, the optimization performed in purpose to obtain the core configuration of KSNP-1000 which offering longest cycle length and flattest power distribution with a constraint on fixed amount and type of KSNP-1000's fuel assemblies.

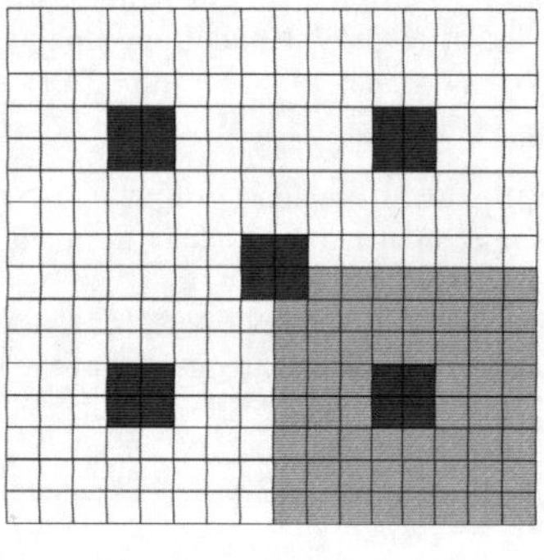

Fig. 3. Fuel assemblies' homogenization to be used in SRAC.

2. Method

2.1 Standard Model Design

Before the computation was done, a standard model of the reactor core had to be created. The standard model is a simplified and homogenized of actual reactor core containing the actual amount of fuel rods and fuel assemblies. The standard model for the works is based on KSNP-1000 from the collaboration document of BATAN and KEPCEO on 1997 [10].

The reactor standard model, in SRAC, was made using PIJ module for fuel rods and fuel assemblies and CITATION module for the reactor core configuration [11]. An example of homogenized fuel assemblies for KSNP-1000 standard model is illustrated in Fig. 3.

Table 1. Fuel Assemblies' Types and Specifications

No.	Type	Number	Enrichment (w/o)	Number of *Gd-rod*
1	A	45	<1.5	-
2	B	20	<2.5	-
3	B1	8	<2.5/1.5	8
4	B2	16	<2.5	4
5	C	12	<3.0/2.5	-
6	C1	32	<3.0/2.5	8
7	D	12	<3.5/3.0	-
8	D1	8	<3.5/3.0	8
9	D2	24	<3.5/3.0	8

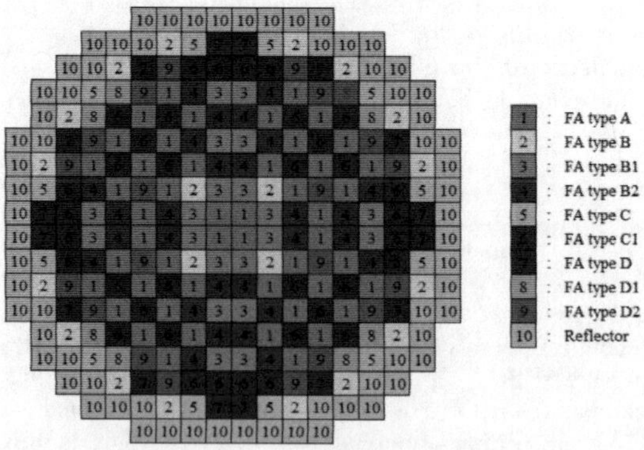

Fig. 4. Standard reactor core model of KSNP-1000.

By designing all fuel assemblies' models and then homogenizing them, the standard reactor core of KSNP-1000 can be built as shown in Fig. 4. The numbers of 1-9 indicate fuel assemblies types differentiated in enrichment as represented on Table I. After core modeling has been done, an adaptation of optimization problems using QEA

https://doi.org/10.1051/e3sconf/20184201007

then can be made. QEA adaptation processes are divided into three steps, i.e., encoding step, processing step and decoding step.

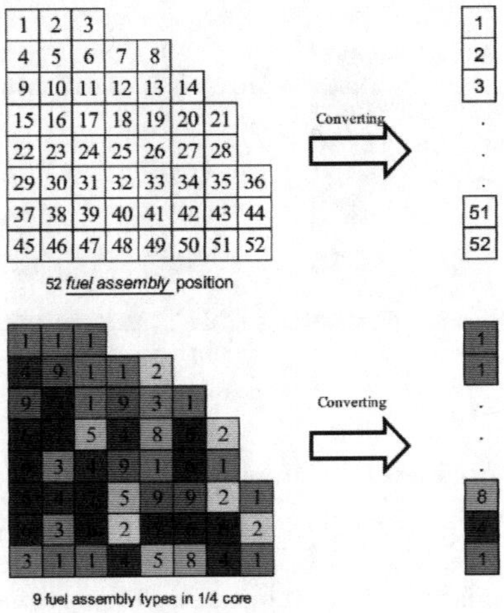

Fig. 5. Core encoding in number series.

In the encoding step, the configuration is encoded in a number series as illustrated in Fig. 5. Then, fuel assemblies adaptation in QEA is used. In this case, the Q-bit individual that had a length of 8 Q-bits was used as appeared on Fig. 6.

The next step is the processing step. In this step, Q-bit number series will be observed as corresponding binary numbers. As the number of Q-bits used is eight, the binary numbers formed is eight bits as well.

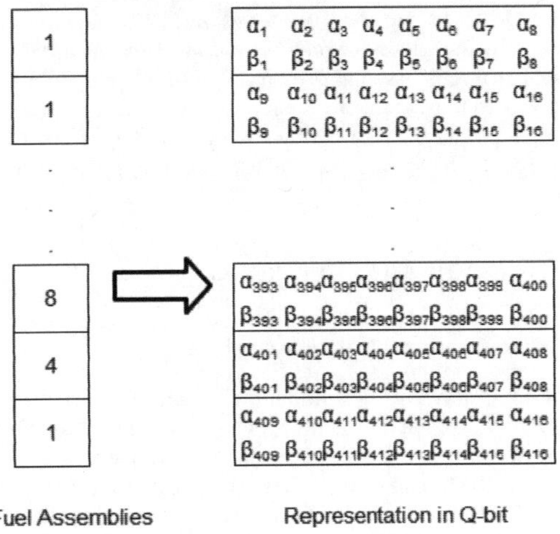

Fig. 6. Problem representation in Q-bit Optimization Steps.

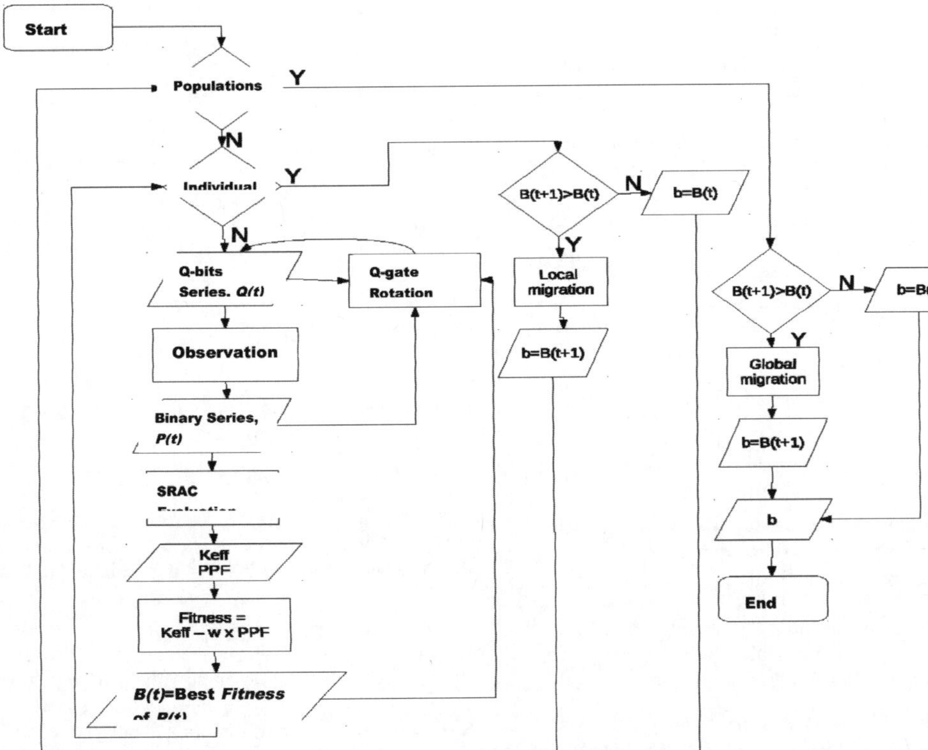

Fig. 7. Flowchart of QEA procedure

A fuel assembly can be represented as a weight value of assembly order, *v*. The value of *v* is used as a weight to sort the fuel assemblies' order.

In a generation, each population's *b* value will be compared. If the value is greater, then global migration happens and the *b* value is replaced by the highest value.

The decoding step process is a reverse of encoding step. In this step, conversion is done from Q-bit series back to fuel assemblies. Firstly, Q-bits pair series is observed to be eight bits binary number series. The binary numbers then are weighted proportionally to certain integers. The flowchart of QEA procedure is illustrated in Fig. 7

3. Result and Discussion

3.1 Determining Rotation Angle, Δθ

In the Δθ value determination to suit the optimation problem, population numbers of 10 and termination parameter, γ_0=0.9 were used. The $\Delta\theta$ variation used are 0.01π, 0.02π and 0.05π. The optimization program was run for 30 times, and the best, mean and worst value were searched. Repetition amount of 30 times minimum is suggested to get best, mean and worst value which is statistically fine [5].

Based on Fig. 8, mean *fitness* evolution ratio can be achieved by each $\Delta\theta$ variation. The highest fitness from $\Delta\theta$=0.01π is 1.107, obtained after 260 generations. This is a too-large generation to achieve convergence. On the other hand, for $\Delta\theta$=0.02π, the convergence generation is only 160 and for $\Delta\theta$=0.05π was far shorter, i.e., 60 generations.

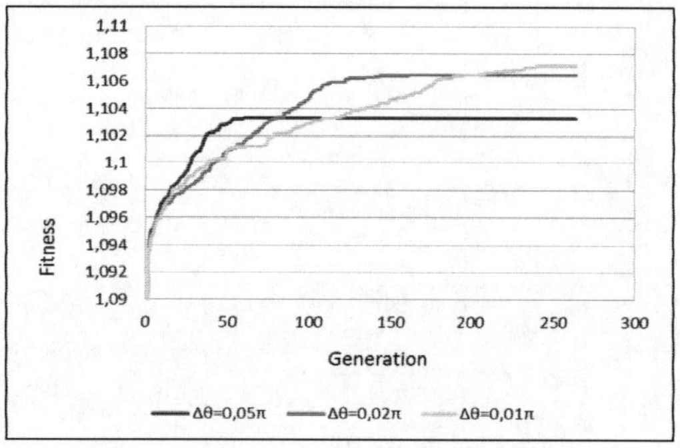

Fig. 8. Evolution of mean *fitness* for each Δθ variations.

In addition, to describe the comparison of each variation, Fig. 9 compares the best *fitnesses* when convergence was achieved for each variation.

The chosen $\Delta\theta$ indeed must have the highest *fitness*. However, based on Fig. 8, the $\Delta\theta$=0.01π needs 260 generations to converge, twice as large as needed for $\Delta\theta$=0.02π. Fig. 9 also shows that rotation angle of 0.01π offered only approximately 0.03% improvement of the best *fitness* than that of 0.02π. Based on such consideration, the parameter $\Delta\theta$=0.02π is chosen for doing the optimization.

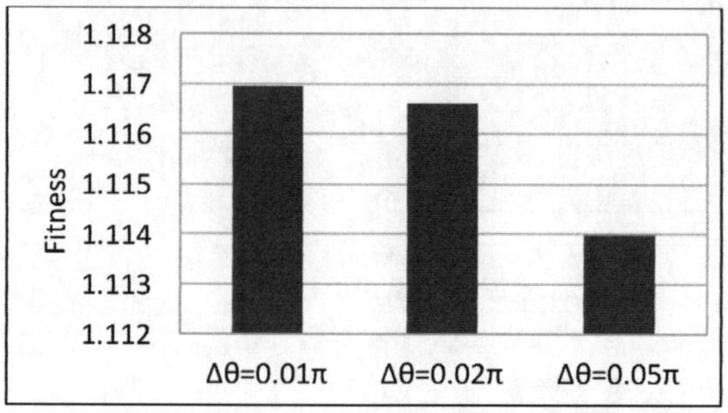

Fig. 9. Best *fitness* value for each rotation angle.

3.2 Determining Weight Factor, w

Determining weight factor was conducted by running the optimization program for 30 times for each value of *w*. The *w* variations were determined on 0.01 up to 0.04 with 0.01 increment. Rotation angle, $\Delta\theta$, used in the optimization was 0.02π which is the optimum value obtained from the previous calculation. A plot of *PPF* on each *w* values can be seen in Fig. 10.

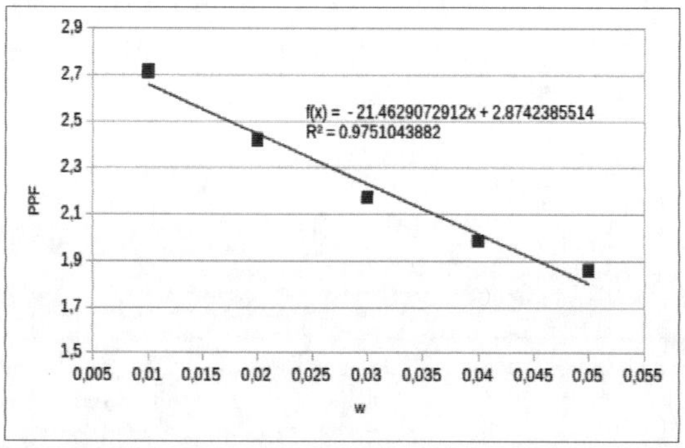

Fig. 10. *PPF* as a function of *w*.

To obtain *w* on $PPF_{standard}$, inversion must be done to the regression equation. Thus, the obtained result can be described as equation (6):

$$w = (-0.04659)\times PPF + 0.13392 \tag{6}$$

By including PPF=$PPF_{standard}$=2, obtained *w* is 0.04074. For an anticipation reason of PPF increasing on BoC (Beginning of Cycle), *w* is then rounded to 0.041.

3.3 Optimized Core

By using Q-gate rotation angle value, $\Delta\theta$=0.02π and the weight factor, *w*=0.041, optimization calculation of KSNP-1000 fuel loading pattern had been conducted. The number of populations used was 20, twice of those on determining the angle and weight factor calculations.

 https://doi.org/10.1051/e3sconf/20184201007

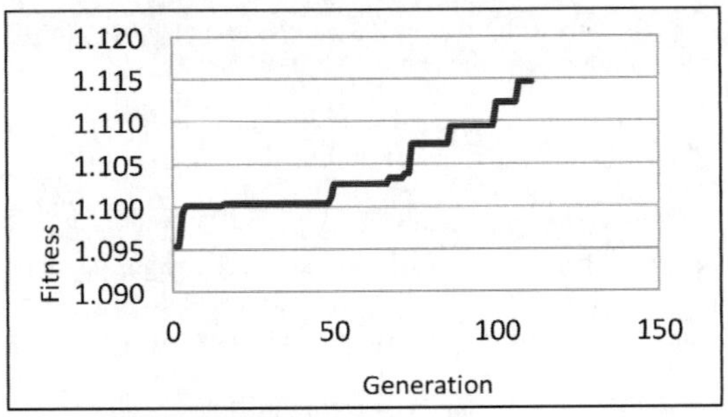

Fig. 11. Fitness evolution on each generation.

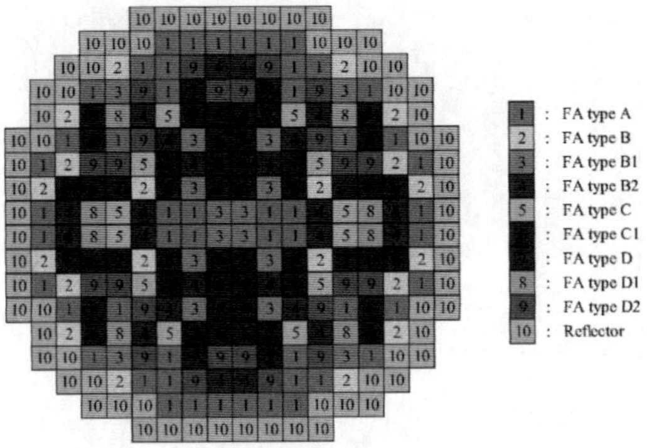

Fig. 12. Optimized core configuration.

From the optimization result, the best *fitness* obtained based on the convergence criteria is 1.11464. The k_{eff} and PPF value after 360 days of burn-up periods respectively are 1.11233 and 1.94367. *Fitness* evolution on various generation can be seen in Fig. 11 and the optimized core configuration is illustrated in Fig. 12.

Thirty samples obtained from optimization calculation are sorted on each fitness value as shown in Table 2. However, the fitness value is not the most prioritized in choosing optimization core, but also PPF must be considered. The standard operation PPF of KSNP-1000 reactor is 2. Therefore, the chosen core configuration must have PPF under the value of 2 [7]. On Table 2, core configuration with the best *fitness* has PPF value larger than 2, which is 2.11019. Because of aforementioned reason, the second best core configuration was then chosen.

Table 2. Optimization Result Sorted By Fitness

No.	*Fitness*	k_{eff}	*PPF*
1	1.11577	1.12029	2.11019
2	1.11464	1.11233	1.94367
3	1.11436	1.11114	1.92135
4	1.11427	1.11681	2.06203
5	1.11403	1.10524	1.78546
6	1.11182	1.11191	2.00205
7	1.11149	1.10782	1.91050
8	1.11090	1.10129	1.76568
9	1.10941	1.10647	1.92836
10	1.10919	1.10762	1.96155
11	1.10879	1.10901	2.00523
12	1.10872	1.10900	2.00684
13	1.10802	1.11249	2.10893
14	1.10793	1.10016	1.81065
15	1.10778	1.10390	1.90539
16	1.10765	1.10801	2.00874
17	1.10744	1.10146	1.85435
18	1.10696	1.10569	1.96919
19	1.10662	1.10866	2.04990
20	1.10604	1.11006	2.09807
21	1.10599	1.11220	2.15135
22	1.10584	1.10710	2.03077
23	1.10584	1.10600	2.00397
24	1.10568	1.10068	1.87796
25	1.10485	1.10995	2.12423
26	1.10472	1.10712	2.05851
27	1.10321	1.10349	2.00684
28	1.10299	1.09887	1.89964
29	1.10197	1.10428	2.05628
30	1.10193	1.10320	2.03108

3.4 Comparison of The Optimized and The Standard Core

The calculation result of fuel loading pattern optimization should be compared with the standard configuration's parameters. Optimization result and standard fuel loading pattern had been tested on a various burn-up level by 60 EFPDs (Equivalent Full Power Days) interval. The value of k_{eff} on each step and PPF is detailed in Table 3.

Based on Table 3, it seems that k_{eff} of the optimized core is always higher than the standard one. However, a high k_{eff} commonly implies to a high PPF as well. Based on the calculation of PPF at BOC, optimized fuel loading pattern is 2.19125, slightly higher than the standard operation of KSNP-1000, which is 2. Also, it is higher than that of the standard core, i.e., 1.50983. The changes of k_{eff} on each burn-up step can be seen in Fig. 13.

Table 3. Standard and Optimized Core's Keff and PPF

EFPD	Optimized		Standard	
	k_{eff}	*PPF*	k_{eff}	*PPF*
0	1.26905	2.191	1.21210	1.510
60	1.22895	2.004	1.18239	1.447
120	1.21394	1.994	1.16869	1.643
180	1.18718	1.955	1.14202	1.661
240	1.16108	1.961	1.11612	1.666
300	1.13616	1.966	1.09150	1.667
360	1.11233	1.944	1.06808	1.663
420	1.08943	1.980	1.04569	1.659
480	1.06732	1.987	1.02422	1.644
540	1.04590	1.992	1.00538	1.624
600	1.02512	1.995	0.98373	1.598
660	1.00496	1.996	0.96478	1.608
720	0.98540	1.994	0.94568	1.653

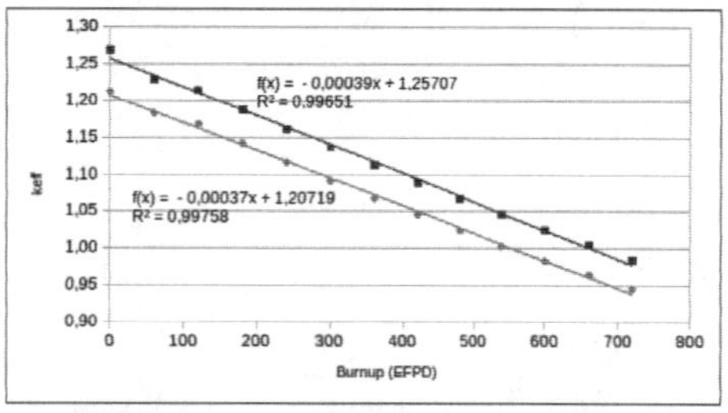

Fig. 13. Changes of k_{eff} on each burn-up level.

To obtain the cycle length on the standard model, first, inversion on regression function of the standard model must be done to get the burn-up function in k_{eff}. The relation between k_{eff} and burnup can be written as

$$burn\ up = (-2702.7027) \times k_{eff} + 3262.67568 \quad (7)$$

By setting *keff* equals to 1, then we have the cycle length of the standard configuration of 560 EFPDs.

Using the same way, the optimized core regression function is inverted to have a burn-up function in k_{eff} like that of Equation (8).

$$burn\ up = (-2564.10256) \times k_{eff} + 3223.2564 \quad (8)$$

Again, by setting *keff* equals to1, the obtained cycle length of the optimized fuel loading pattern is 659 EFPDs. Relative comparison of cycle length and PPF at BOC for the standard model and the optimized core is shown in Fig. 14.

Optimized core configuration results in the reactor's cycle length to be 17.67% longer than that of the standard core. The longer cycle length will be able to extend the operation time between refuelings. Thus, it will economically increase the profit of the nuclear power plant. The longer cycle length also affects the cost of nuclear fuels to be decreased as fuels can stay longer in the reactor core in the same operation power.

On the safety aspect, PPF of the optimized core is 45.13% higher than that of the standard model of KSNP-1000. The value of PPF on BoC is 2.191. The increasing in PPF can affect power distribution of optimized core to be worse than the standard core. However, slightly higher than 2 of PPF are still can be accepted as core configuration of KSNP-1000 [11]. The difference of power density distribution between the standard core and the optimized core can be seen in Fig. 15.

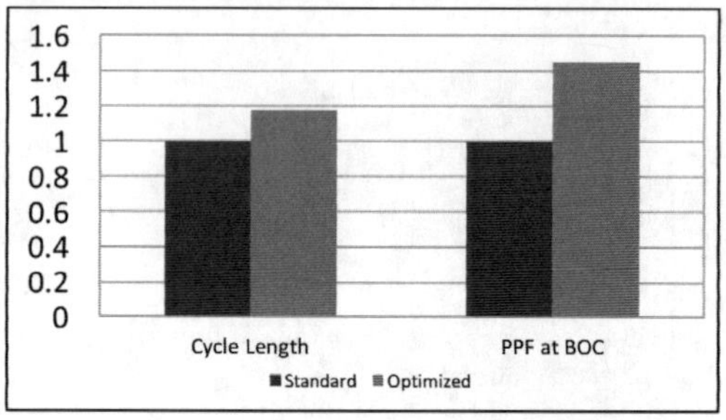

Fig. 14. Relative comparison on cycle length and PPF on BoC.

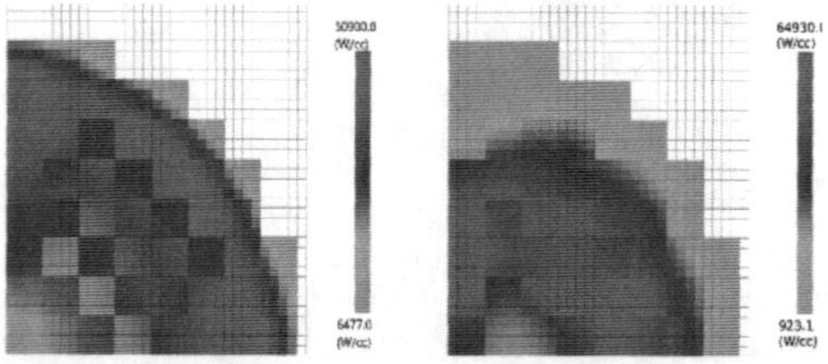

Fig. 15. Power density distribution of standad core (left) and optimized core (right).

Average power is also a parameter need to be calculated. Ideally, the optimized and standard core must have the same power. Average power can be calculated based on Equation (9).

$$Average\ Power = \frac{Maximum\ Power}{PPF} \quad (9)$$

Because of the same core volume between the optimized and standard core, the power comparison can be computed from the power density with the unit of W/cc. By using the maximum power density value in Fig. 15 and PPF on Table 3, obtained power density of optimized and standard core respectively are 29.63 kW/cc and 33.71 kW/cc.

4. Conclusion

Using rotation angle value $\Delta\theta$ equals to 0.02π and weight factor, *w* equals to 0.041, an optimized core configuration can be obtained, resulting 17.67% longer in fuel cycle than that of the standard core. Also, the PPF can be slightly maintained on the value of 2. Therefore, Quantum-inspired Evolutionary Algorithm (QEA) can be used to optimize fuel loading pattern effectively.

References

1. D.G. Cacuci, *Handbook of Nuclear Engineering Volume I: Nuclear Engineering Fundamentals* (Springer, New York, 2010)
2. J. Lewins, M. Becker, *Advances in Nuclear Science and Technology* (Kluwer Academic Publishers, New York, 2002)
3. A.D.S. Nicolau, R. Schirru, A.M.M.D. Lima, Prog. Nucl. Energ **55**, 40-48 (2012)
4. K.H. Han, J.H. Kim, IEEE Transactions On Evolutionary Computation **6**, 580-593 (2002)
5. K.H. Han, J-H. Kim, IEEE Transactions On Evolutionary Computation **8**, 156-169 (2004)
6. Z. Laboudi S. Chikhi, Int. Arab. J. Inf. Techn **9**, 243-249 (2012)
7. M.H.D. Silva, R. Schirru, Prog. Nucl. Energ **74**, 103-109 (2014)
8. D. Hartanto, *Efek Pemodelan Sel Bahan Bakar Pressurized Water Reactor Hipotesis terhadap Nilai Perlipatan Tak Hingga [The Effect of Cell Modeling of a Hypothetical Pressurized Water Reactor Fuel to Infinite Multiplication Factor].* Internship Report, Department of Engineering Physics, Universitas Gadjah Mada, Yogyakarta, (2006) (in Bahasa Indonesia)
9. F. Arbiyani, *Analisis Kekritisan dan Kebutuhan Bahan Bakar PWR KSNP-1000 Berpengkayaan 4% Sampai 5% [Analysis of Criticality and 4% to 5% Enriched Fuel Requirement in KSNP-1000 PWR].* Undergraduate Thesis, Department of Engineering Physics, Universitas Gadjah Mada, Yogyakarta, (2005) (in Bahasa Indonesia)
10. Y.P. Sitompul, *Optimasi Fuel Loading Pattern Teras PWR menggunakan Metode Genetic Algorithm [Optimization of Fuel Loading Pattern of a PWR Core Using Genetic Algorithm Methods].* Undergraduate Thesis, Department of Engineering Physics, Universitas Gadjah Mada, Yogyakarta, (2008) (in Bahasa Indonesia)
11. K. Okumura, *A Comprehensive Neutronics Calculation Code System* (Japan Energy Atomic Agency, Ibaraki, 2007)

 https://doi.org/10.1051/e3sconf/20184201008

Design of Hydro Power by Using Turbines Kaplan on The Discharge Channel Paiton 1 and 2

Alvin K. Sosilo, Harsono Hadi, and Totok Soehartanto

Engineering Physics Dept. Institut Teknologi Sepuluh Nopember, Surabaya, Indonesia

Abstract..Condenser water from the discharge channel PJB Paiton discharged to the sea has the potential mechanical energy, because the flow rate of 7.6 m3 / s (if both discharge PJB Paiton function) and the discharge channel reaches a height of 4m. This paper will describe the design of hydro power (in the form of a block diagram) by using Kaplan turbine driven by utilizing the wastewater condenser. Kaplan turbine performance represented in the form of the relationship between the incoming water flow and the pitch angle (the angle between the propellers with a hub) to the torque generated. The simulation results indicate that the turbine torque is proportional to the mechanical power turbine. The greater the torque, the greater the mechanical power, and vice versa.

1 Introduction

In the steam power plant contained equipment such as boilers, condensers, turbines, generators, and pumps. The condenser is a heat exchanger, comprising the cooling water used to cool the steam turbine output. The steam will be condensed into water condensate, and condensate water will be fed back into the boiler using a pump (Syamsurijal, 2009). Thus, the cycle of the steam power plant is a closed cycle or cycles that do not require additional water during a sufficient condition. During operation, the condenser needs water as a cooling medium steam through heat exchange scheme.

After cooling the steam, the water will be discharged through a channel called discharge channel. Until now there has been no initiative to harness the water discharge channel's output, there is potential for enormous energy (discharge and water level discharge channel output of about 7.6 m^3/s and 4m). This condition has the potential to be used as a system of Hydroelectric Power Plant (HPP).

Water turbine consists of several types, determination of water turbines in hydropower system design based on the characteristics of water flow (flow rates and falling stream). One type of water turbine is a Kaplan turbine. The pitch angle of the turbine Kaplan is the angle between the propeller with a hub. Variations in pitch angle and variations in water discharge, used for calculation of torque.

2 Basic Theory

2.1 Condenser

Condenser at hydroelectric plants is a heat exchanger, serves to cool the steam turbine output. The water condensation steam turbine output is called condensate water. Condensate water is the main source of water supply to the power plant boiler filler. Process on the steam power plant is a closed cycle; this is due to water vapor condensation turbine output is used again in a process in the boiler (Fig. 1).

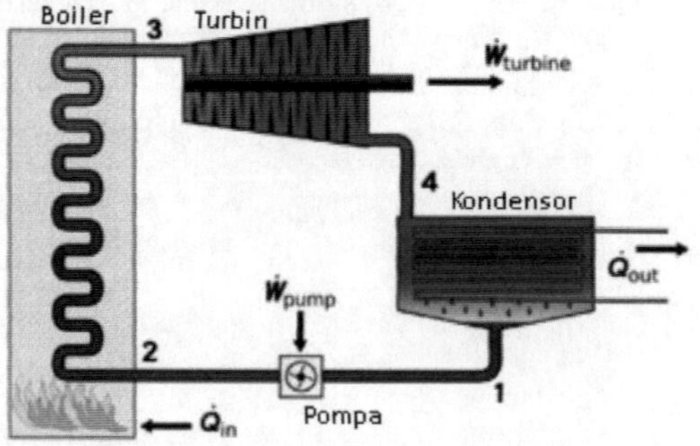

Fig. 1. Hydro power plant Cycle (Jumawa, C., 2013

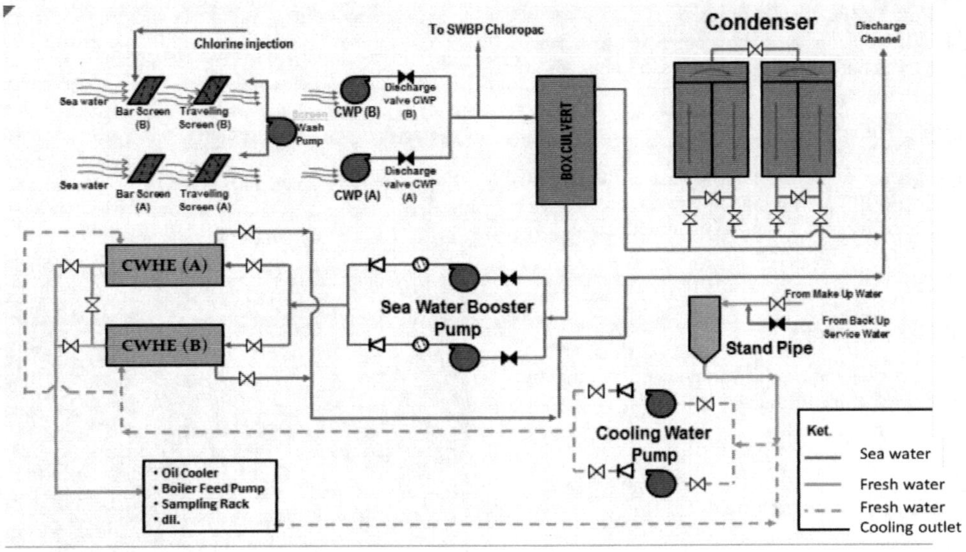

Fig. 2. Schema of discharge channel of HPP (Farzi, A, 2015)

2.2 Discharge channel

Conditioning water of condenser PJB Paiton use sea water (has undergone several processes, such as bar screen, traveling screen, screen wash pump, cooling water pump (CWP), water coolers are housed in box culvert, and then discharged as a cooling medium in the condenser. Conditioning water will be out towards the discharge channel (to dispose of condenser cooling water to the river). Here is a scheme of water flow in the discharge channel of steam power plants.

2.3 Water Turbin Characteristic

Water turbine has the characteristics stated in general by 6 pieces constants, among others (Luknanto, 2013):

2.3.1 Speed Ratio (ϕ)

The speed ratio is the ratio between the translational speed of the turbine to the speed of the water through a nozzle with an altitude plunge H. Rated speed ratio is proportional to the height H of waterfalls on the turbine.

$$\phi = \frac{V_{linier}}{\sqrt{2gH}} = \frac{\frac{N\pi D}{60}}{\sqrt{2gH}} \tag{1}$$

N = turbine rotational speed (rpm)
D = diameter turbine(m)
H = effective high waterfall / head(m)

2.3.2 Unity speed

Unity turbine speed is the rotational speed as a function of the diameter of the turbine and high waterfalls from the discharge channel.

$$N = 84{,}6\phi\frac{\sqrt{H}}{D}$$
$$N_U = 84{,}6\phi \tag{2}$$

2.3.3 Unity discharge

Unity discharge that goes to the turbine as a function of the diameter of the turbine and high-plunge from the discharge channel.

$$Q_U = C_d\frac{1}{4}\pi\sqrt{2g} = \frac{Q}{D^2\sqrt{H}} \tag{3}$$

Q = water discharge on turbines(m^3/s)
C_d= coefficient of discharge

2.3.4 Unity power

Turbine power generated proportional to the diameter and height of the discharge channel plunge.

$$P_U = \eta\rho g Q_U$$
$$= \frac{P}{D^2H^{3/2}} \tag{4}$$

P = mechanical power turbine(watt)
ρ = the density of water(kg/m^3)
g = acceleration of gravity(m/s^2)
η = turbine efficiency

2.3.5 Diameter Specification

Diameter turbines are needed to produce one unit of power at high waterfall H.

$$D_s = \frac{1}{\sqrt{P_U}} = \frac{DH^{3/4}}{\sqrt{P}} \tag{5}$$

2.3.6 Specific Speed

Rotational speed turbines to produce one unit of power at high waterfall H.

https://doi.org/10.1051/e3sconf/20184201008

$$= \frac{N\sqrt{P}}{H^{5/4}} N_S = \sqrt{P_U} N_U \quad (6)$$

In addition to the six constants above, there are also other important characteristics in water turbines, namely:

2.3.7 Torque Turbine

The energy generated by the turbine rotates due to the flow of water takes. Water turbine torque can be obtained from the equation of impulse and momentum [3].

$$dF_{a/p} = \frac{dmV_{rel}}{dt} \quad (7)$$

2.3.8 Turbine efficiency

Comparison between the mechanical power generated by the turbine to power the turbine water flow [4]

$$= \frac{\tau\omega}{\rho QgH} \eta = \frac{P}{\rho QgH} \quad (8)$$

2.4 Selection of water turbines

There are some parameters in determining the types of water turbines, among others (Patty, 1995):

2.4.1 High effective waterfall (head) and discharge that will be used to drive a turbine.

In Fig. 3 shows the graph turbine type selection based on the height H and discharge.

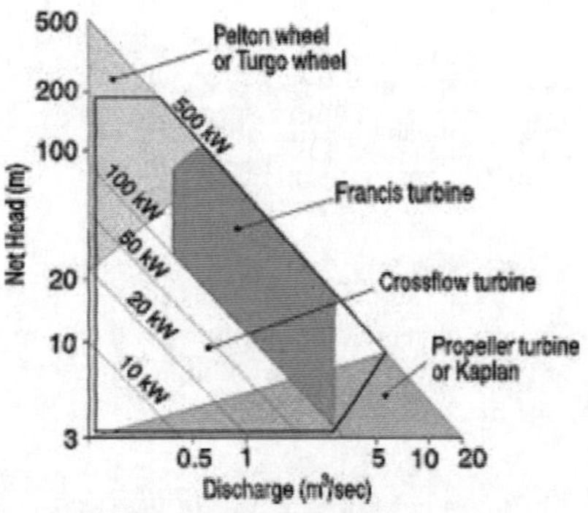

Fig. 3. Types of turbines based head & discharge

Pelton turbine is suitable for high head, and low flow while effectively Kaplan turbines for low head and flow profusely. Francis turbines and crossflow turbines for head and low flow-medium.

2.4.2 Specific Speed Turbine

The value range of specific speed turbine water:

- $N_s = 9 - 25$ (Pelton turbine with one jet)
- $N_s = 25 - 60$ (Pelton turbine with more than one jet)
- $N_s = 40 - 400$ (Francis turbine)
- $N_s = 260 - 860$ (Kaplan turbine)

Specific Kaplan turbine speed can be calculated using equation

$$N_S = \frac{9431}{H+9{,}75} + 155 \quad (9)$$

3 Hydro power design

In this study, Kaplan turbines will be placed under the discharge channel as shown in Fig. 4.

Fig. 4. Position Kaplan turbine

Installation of the system is still not there, then the channel discharge channel contained in steam power plants have to be modified in order Kaplan turbine placement can be done well.

3.1 Block Diagram of hydro power

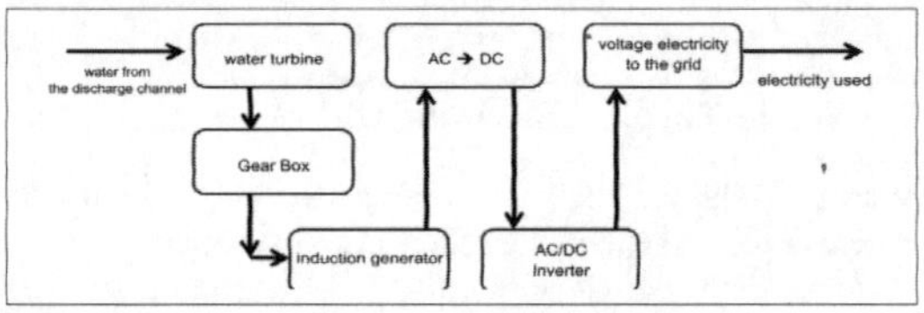

Fig. 5. Block diagram of hydro power system

Water discharge channel output will hit Kaplan turbine so that the turbine rotates. Turbine connected to a generator rotor through a gearbox. rotor spinning will make the generator produces alternating electrical voltage. This electrical voltage is rectified because the energy reservoir hydropower system is erratic. voltage electricity supplied is alternating electrical voltage, the voltage that has been rectified converted back to an alternating electrical voltage by a DC / AC Inverter before being connected to a voltage mesh nets.

This paper represents only the components of the turbine, gearbox, and generator. discussion after the generator components are recommendations that can be given for the reasons mentioned above.

https://doi.org/10.1051/e3sconf/20184201008

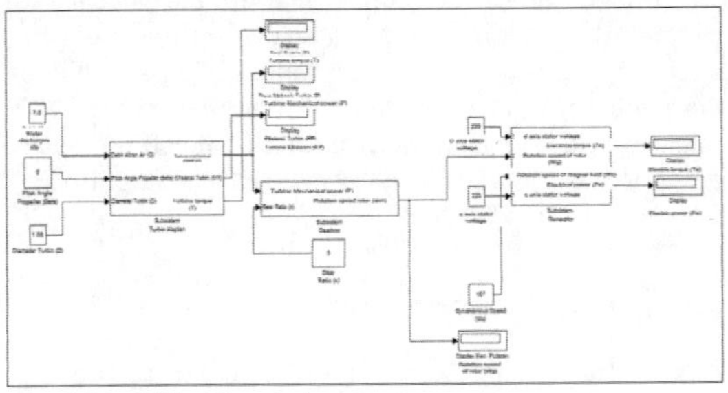

Fig. 6. Modelling the whole system

The system consists of three major subsystems, namely the water turbine, gearbox, and generator induction.

Input system:

- Water flow rate (Q), (m^3/s)
- *pitch angle* (β), (degree)
- turbine diameter (D), (m)
- *gear ratio* (n)

Output system:

- turbine torque (τ), (Nm)
- mechanical power of turbine (P), (watt)
- turbine efficiency (%)
- the rotated speed of rotor generator (ω_g), (rad/s)
- electrical torque (T_e), (Nm)
- electrical power (P_e), (watt)

3.2 Hydro turbine block system

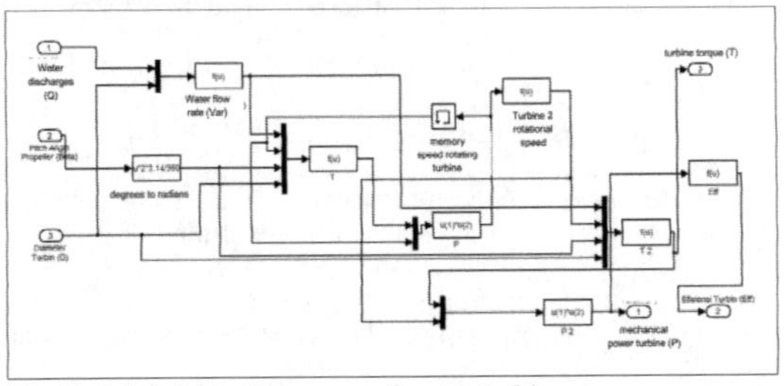

Fig. 7. Modeling sub-system of water turbines

Input:

- Water flow rate(Q), (m^3/s)
- *pitch angle* (β), (degree)
- turbine diameter (D), (meter)

Output:

- turbine torque (τ), (Nm)
- mechanical power of turbine (P), (watt)
- turbine efficiency (%)

3.3 Determining of turbine type

Turbine type is determined through the selection of a water turbine curves (Fig. 8).

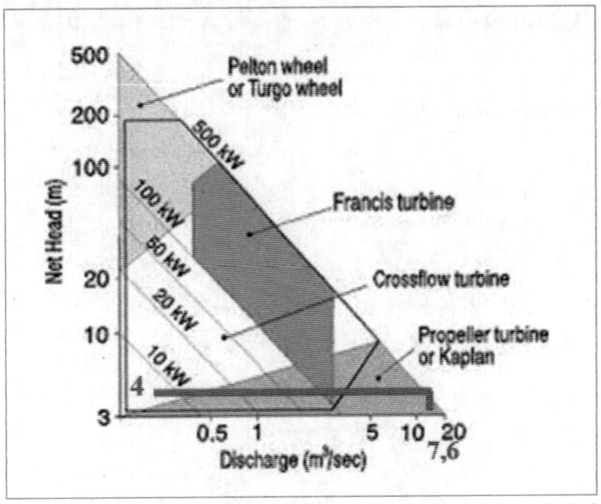

Fig. 8. Determination the type of turbine based on flow rate and altitude

Through the Fig. 8, it was found that the type of turbine that is suitable for the conditions of the discharge channel PJB Paiton is Kaplan turbines or propellers.

The determination of the specific speed (NS) using Equation 6 (for the height H = 4m):

$$N_S = \frac{9431}{H + 9{,}75} + 155 \\ = 840{,}89 \tag{10}$$

NS value calculation results in accordance with NS value for Kaplan turbines (in the range 260-860).

3.4 Determination of turbine dimension

Dimensions calculated based on the Kaplan turbine propeller diameter and number of propellers. wicket gate which is part of the steering flow of water, its dimensions are based on previous research [6]. Wicket gate opening angle of 30^0

Discharge water before entering and after exiting the wicket gate assumed to be the same so that the propeller diameter (D) defined by the equation 5, namely:

$$D = cQ^{0{,}5} \tag{11}$$

C value is determined based on Table 1, was obtained: c = 0.39 for the value of H in the range of 5 m.

Table 1. Determination coefficient value c [1]

H (m)	5	10	15	20	25
C	0,39	0,37	0,37	0,36	0,36

By entering a value of Q = 7.6 m^3/s then obtained:
= 1.08m

The number of the propeller (Z) is determined by Table 2 (based on N_s values have been obtained).
Z = 4 for NS value in the range of 800.

 https://doi.org/10.1051/e3sconf/20184201008

Table 2. Determination of Number of Turbine Blade Runner (Z)

Ns (Hp)	400	500	600	700	800
Ns (kW) = 0,857Ns	340	430	510	600	680
Z	8-6	7-6	6-5	5-4	4

Kaplan turbines are used to have the initial specification as follows:

Diameter (D) = 1,08 m,
Number of propeller (Z) = 4 unit,
*wicket gate*opening = 30^0.

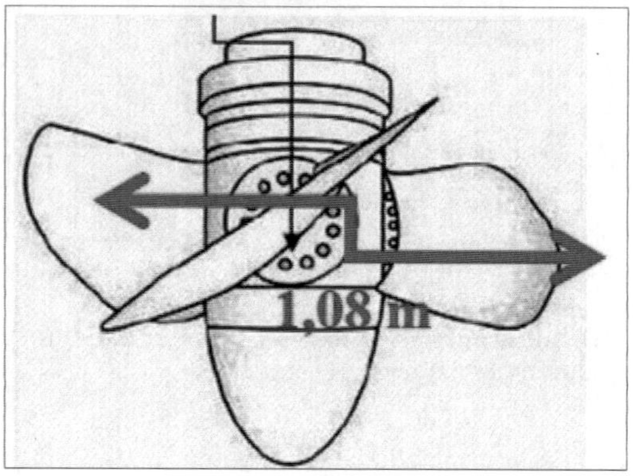

Fig. 9. Part propeller turbines Kaplan

3.5 Determining gear box parameter

Parameter gearbox is required as the coefficient convert the turbine to the generator output. In this study, the parameters used in the gearbox is the gear ratio (n).

Gear ratios will be used as one input parameter and varied. Gear ratio according to Kaplan turbines ranges from 1: 3 to 1: 7 (http://directindustry.com). N input values determined in the modeling with:

$n_1 = 5$; $n_2 = 6$; and $n_3 = 7$

4 Analysis And Discussion

Pitch Angle Torque relationship with Power Turbine and Turbine Mechanics.

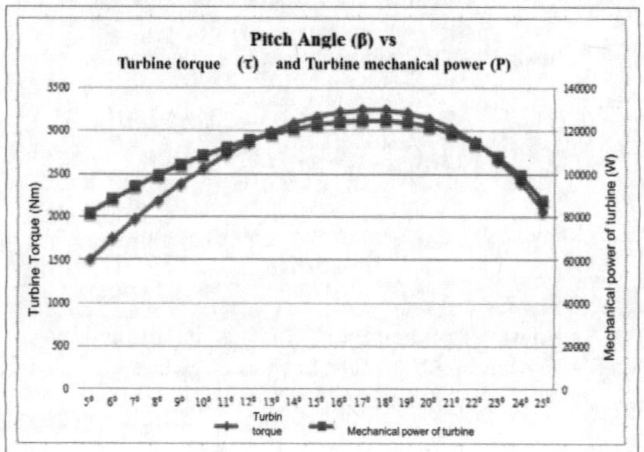

Fig. 10. Graph relationship with the pitch angle torque and mechanical power turbine

Turbine torque is proportional to the mechanical power turbine. large torque obtained in the range of pitch angle 15^0 - 20^0 with maximum values of 3249.7 Nm on the pitch angle 17^0. torque of the turbine will have an impact on mechanical power turbines are getting bigger, the pitch angle range will generate maximum value 124,987.1 W or 125 kW.

Acknowledgment

Acknowledgements to PJB Paiton that provides opportunities for writers to do the design of hydropower by utilizing the condensate flow from the discharge channel.

References

1. I. Akbar, Studi Kelayakan dan Penyusunan Basic Design Proyek PLTA Siteki. Vol. 1 Bab 3 "Parameter Design Optimum" (2003)
2. S. Chapman, *Electric Machinery Fundamentals 4th edition* (2005)
3. A. Crawford, J. Motley, Modeling Torque of a Hydro-Turbine (2011)
4. M. Gustafsson, Improved Governing of Kaplan Turbine Hydropower Plants Operating Island Grids (2013)
5. http://engineeringtoolbox.com/electrical-motors-hp-torque-rpm-d_1503.html diakses 13 Januari 2016 pukul 17.00.
6. A. Susanto, Perancangan dan Pengujian Turbin Kaplan pada Ketinggian (H) 4 m Sudut Sudut Pengarah 30^0 dengan Variabel Perubahan Debit (Q) dan Sudut Sudu Jalan. Jurusan Teknik Mesin Fakultas Teknik. Universitas Muhammadiyah. Surakarta, 2013.
7. A. Farzi, Sistem Pendingin. PT. PJB UP Gresik. (2015)
8. C. Jumawa, Pembangkit Listrik Tenaga Uap. Group Riset CFD Indonesia. Ahmad Indra Research. Universitas Indonesia. Depok (2013)
9. D. Luknanto, Bangunan Tenaga Air. Jurusan Teknik Sipil Fakultas Teknik. Universitas Gadjah Mada. Yogyakarta (2013)
10. O. Patty, Tenaga Air. Surabaya: Erlangga (1995)

Potential Analysis of Ex-Coal Mining Land As Pumped Storage Hydro Powerplant In Kutai Kartanegara, East Kalimantan

Yollanda Z. Devi[1,*], *Andang* W. Harto[1], *Rachmawan* Budiarto[1], and *Novi* Trihastuti[2]

[1]Department of Nuclear Engineering and Engineering Physics, Universitas Gadjah Mada, Yogyakarta, Indonesia
[2]PT. ASI Pudjiastuti Geosurvey, Jakarta Indonesia
*yollandazd@gmail.com

Abstract. The growth of electricity demand will increase 56.8% in 10 years. Meanwhile, the fulfillment of electricity in Indonesia is dominated by coal. The amount of electricity growth has not balanced yet by the amount of electricity supply. It causes the electricity deficit in some areas, especially at the peak load period. One of the areas that often have electrical power deficit is East Kalimantan, although the province is one of the biggest coal mining areas in Indonesia. The mining activities give several impacts in environmental damages especially in the ex-coal mining land which has not been well managed. The impact includes an irreversible change of landscape and huge craters caused by open mining activities. In this study, an analysis is conducted for utilization of ex-coal mining land in Kutai Kartanegara as a reservoir of energy storage system type Pumped Hydroelectric Storage (PSH). The analysis provides a comparison of three variations of the piping line system. The study uses the spatial data in the form of Digital Elevation Model (DEM) to calculate the time of starting operation system, expected generated energy, required energy for power plant operation and the efficiency of the system. This study compares some criteria based on three variations of the piping line system and the effect on the efficiency of the system. **Keywords— Pumped Storage Hydro Powerplant, starting operation time, energy turbine, energy of pump, efficiency**

1 Introduction

The electricity demand continuously increases until 56.8% in 10 years, from 637.7 kWh/person in 2010 to 999.9 kWh/person in 2020 [1]. The growth of electricity demand cannot be balanced by the growth of electricity supply yet. It causes electrical power deficit in some areas, especially in the peak load period. One of the electrical networks that often have electrical power deficit is Mahakam System.

In the other hand, an area covered the Mahakam System is Kutai Kartanegara which has many coal mining activities. The mining activities there causes the environmental damage. Most of the mining areas were originally forest area which became the ex-coal mining land [2]. The coal mining process which uses the open mining technique with dig and fills method but does not provide proper reclamation and revegetation will cause the formation of huge craters. The formation of huge craters that represents irreversible environment damage gives the opportunity to be transformed as Pumped Storage Hydro Powerplant (PSH). The using of ex-coal mining land as the reservoir of PSH system in Kutai Kartanegara can be a part of the solution to electrical power deficit in Mahakam System electrical network.

The system is a hydropower generation system operated to fulfill the electrical demand, especially in the peak-load period. The system requires two water reservoirs in operation, called upper reservoir and lower reservoir. When the electricity demand is above the base-load, the system will flow the water from the upper reservoir to lower reservoir to rotate the turbine. It is called as a discharging mode. Whereas, when the electricity demand is below the base-load, available energy surplus will be consumed by the PSH system to pump the water from lower reservoir to upper reservoir. It is called as charging mode [3]-[7].

This study analyzes the energy produced, the energy needed, starting time operating system and efficiency of the system of the PSH system using the ex-coal mining land in Sanga-Sanga district, Kutai Kartanegara, East Kalimantan. This study uses GIS data which is gotten from LiDAR aerial remote sensing survey in 2014. This study presents an overview and recommendation of ex-mining land utilization as PSH.

2 Methodology

2.1 Literature Study

After reviewing several relevant kinds of literature, then study conducts collecting data, data processing, and analysis. This study requires the geographical data of two ex-coal mining pits as a potential site being analyzed.

Geographical information is analyzed to determine form and volume of craters. The data provide a basis to estimate expected power generation due to available head and discharge as well as the volume of starter water. The other data needed is the local meteorological data to calculate the addition of water stored annually. It is also used to estimate the starting operation time based on the volume of water in the storage.

Data analysis is conducted in some steps as follow :

- Generating a contour based on map in the form of Digital Elevation Model (DEM) using Global Mapper 10
- Sampling the contour data per 1 m using ArcMap 11
- Calculating the volume and area based on contours using Global Mapper 10
- Implementing regression method to know the function of depth and area also depth and volume
- Estimating additional water in the storage from the local rainfall rate
- Calculating the rate of evapotranspiration and infiltration based on local climate data
- Estimating the net discharge addition of water in the storage
- Estimating the time when upper reservoir being filled until 0 m height of water surface
- Providing the side view contour of both reservoir
- Designing the piping system
- Calculating the energy of pump and turbine energy in a cycle
- Calculating the efficiency of the system

3 Results and Discussion

3.1 Description of Site Study and Pumped Storage Hydro Powerplant Plan Design

The study site is ex-coal mining land which is located in Sanga-Sanga, Kutai Kartanegara, East Kalimantan Province. The study selects two craters with different depth. The coordinate of upper reservoir is 0°41'45" S and 117°14'15" E, and the coordinate of the lower reservoir is 0°40'45" S 117°14'38"E. Fig. 1 is the orthophoto data of ex-coal mining land which is used as the study site.

Both reservoirs in the site have the same height but different depth and volume. The upper reservoir has a smaller volume and lower depth than the lower one. The different depth between two reservoirs is the basic variable to use them as the reservoir of PSH system. Based on the formation of the site, the piping system will be built underground.

3.2 Estimation of Starting Time Operation System

The rainfall is the water source for two ex-coal mining craters used in this study. This PSH system can start to operate when the upper reservoir is filled by rainfall until 0 m height of water surface (see Fig. 1). The study considers the water losses caused by evapotranspiration and percolation. Mathematically, the equation of additional net volume can be written as follow

$$V_{net} = V_r - V_e - V_p \quad (1)$$

$$V_p = v_p \times A_p \times t \quad (2)$$

$$V_e = v_e \times A_e \times t \quad (3)$$

Where :

V_{nett} : net volume (m^3)
V_p : percolation volume (m^3)
V_e : evapotranspiration volume (m^3)
v_p : the rate of percolation (m/year)
v_e : the rate of evapotranspiration (m/year)
A_p : percolation area (m^2)
A_e : evapotranspiration area (m^2)
t : time (year)

Fig. 1. Orthophoto of ex-coal mining land as site study

Based on climate data, the local rainfall rate is about 2.695 m/year [8], percolation rate is 0.73 m/year [8] and evapotranspiration rate is about 0.535 m/year [9].

The time when the upper reservoir is filled by the water can be found by knowing the volume and area of the upper reservoir in each height. Fig. 2 below shows the graph of depth to area of both reservoirs. Fig. 3 shows the graph of depth to volume of both reservoirs.

Based on the data, the net additional volume in a year is about 1,417,570 m^3. The estimation of starting operation time of this PSH system is about 14.7 years later, or 2028. In that time, the height of water surface is -0.087 m for the upper reservoir and -48.002 m for the lower reservoir.

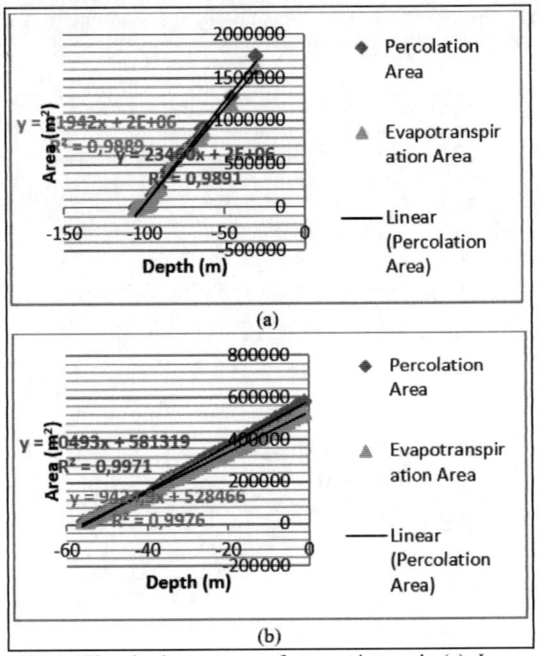

Fig. 2. The depth vs. area of reservoir graph (a) Lower reservoir (b) Upper reservoir

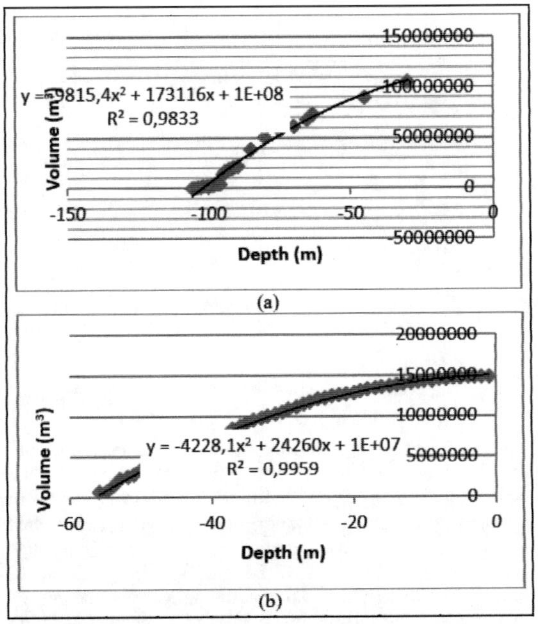

Fig. 3. The graph of depth to volume of reservoir (a) Lower reservoir (b) Upper reservoir

3.3 Pipeline Design

This study uses three variations of the pipeline for Pumped Storage Hydro Powerplant (PSH) system. They are pipeline A, pipeline B, and pipeline C. Fig. 4 shows the three pipeline designs used in this study

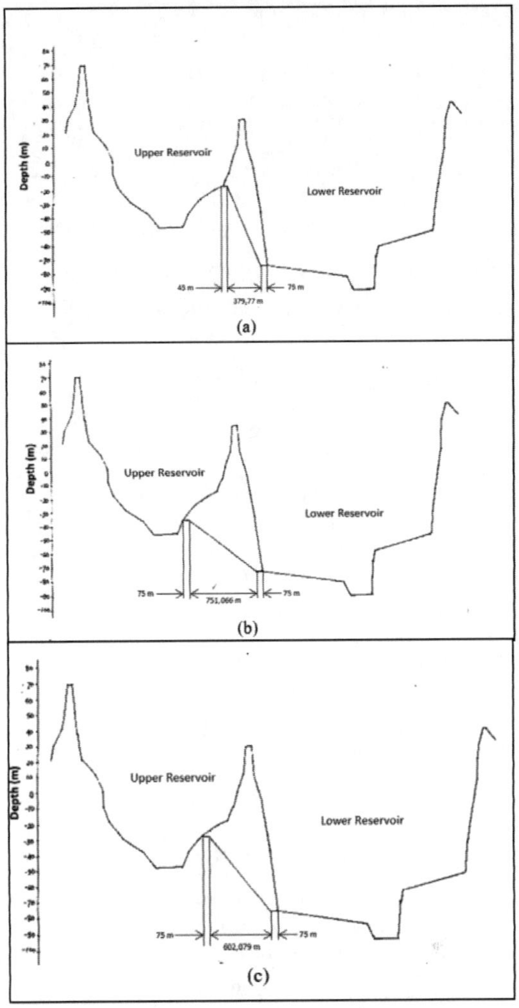

Fig. 4. Layout design of pipeline (a) Design A (b) Design B (c) Design C

The length of pipeline A is about 502.77 m, pipeline B is about 904.066 m, and pipeline C is about 755.079 m. The longer pipeline, the greater head loss will be produced because due to the greater friction. It will influence the amount of power produced by the PSH system. Each pipeline design has 4 m in diameter.

3.4 Energy of Turbine

Therefore, turbine power produced by the PSH system in this study is not constant at the time. It because the flow rate of the water depends on the hydrostatic pressure which declines by the time.The highest power of the system is obtained when the water surface is at the maximum level. The change of flow rate occurs due to the change of water level which causes variation of hydrostatic pressure.

$$\frac{dZ_H}{dt} = \frac{-v.A_p}{\left(A_H(Z_H)+Z_H\frac{dA_H}{dZ_H}\right)} \quad (4)$$

$$\frac{dZ_H}{dt} = \frac{-12.56\sqrt{\frac{2g(Z_H(t)-Z_L(t))}{\left(K+\frac{0.184}{\left(\frac{\rho v s D}{\mu}\right)^{0.2}}\frac{L}{D}\right)}}}{(18849.8\,Z_H(t)+528466)} \quad (5)$$

Where :

Z_H : water level in upper reservoir (m)

Z_L : water level in lower reservoir (m)

g : gravitational acceleration (m/s^2)

ρ : water density (kg/m^3)

v : flow rate (m/s)

D : inner diameter of pipe

μ : water viscosity (kg/ms)

L : length of pipeline (m)

K : friction coefficient

A_H : surface area of upper reservoir water level (m^2)

Equation (4) is applied to calculate the declining rate of water level in the upper reservoir. Substituting variables of the study site into (4) gives (5).

$$\frac{dZ_L}{dt} = \frac{v.A_p}{\left(A_L(Z_L)+Z_L\frac{dA_L}{dZ_L}\right)} \quad (6)$$

$$\frac{dZ_L}{dt} = \frac{12{,}56\sqrt{\frac{2g(Z_H(t)-Z_L(t))}{\left(K+\frac{0{,}184}{\left(\frac{\rho v D}{\mu}\right)^{0{,}2}}\frac{L}{D}\right)}}}{(43884\,Z_L(t)+2000000)} \quad (7)$$

Where

A_L : surface area of lower reservoir water level (m^2)

Furthermore, (6) is applied to calculate the increasing rate of water level in the lower reservoir. Substituting variables representing the study site into (6) gives (7).

$$v^2 + \frac{0.184}{\left(\frac{\rho D}{\mu}\right)^{0.2}}\frac{L}{KD}v^{1.8} - \frac{2g(Z_H-Z_L)}{K} = 0 \quad (8)$$

Then, the flow rate of the water in discharging mode can be found by solving the root of flow rate quadratic (8). Equation (8) can be solved by using Secant method. To find the value of Z_H and Z_L, Runge-Kutta numerical method has to be done with the initial condition $Z_H(0)$ = -0.087 m and $Z_L(0)$ = -48.002 m.

$$P = \frac{1}{2}.\rho.A_p.v^3 \quad (9)$$

Based on the calculation using that equation, the power of turbine can be found by using (9) above

Where :

P : the power generated by turbine (W)

The power of turbine will decline by the time due to the declining of hydrostatic pressure as shown in Fig. 5

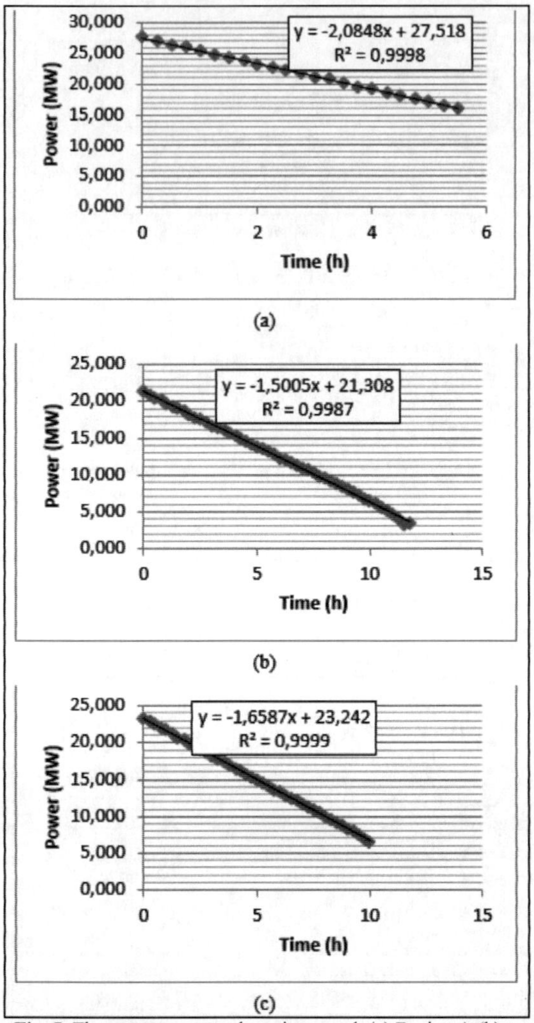

Fig. 5. The power generated vs. time graph (a) Design A (b) Design B (c) Design C

The operation time of turbine per cycle can be done until the water level is 10-15 m above the pipeline because the pipeline has to be submerged in water. 10-15 m is chosen by considering the fluctuation of declining water. Based on the calculation, the operation time of turbine per cycle is 5.5 hours for design A, 12.25 hours for design B and 10 hours for design C. So, mathematically the energy produced by the turbine in a cycle can be written as (10) - (12)

a. Pipeline A

$$E_{turbine} = \int_0^{5.5} -2.0848\,t + 27.518\,dt \quad (10)$$

b. Pipeline B

(11)

$$E_{turbine} = \int_0^{12.25} -1.5106\,t + 21.348\,dt$$

c. Pipeline C

$$E_{turbine} = \int_0^{10} -1.6587\,t + 23.242\,dt \qquad (12)$$

Where :

E : energy produced by turbine (MWh)

Based on those equations, the energy produced by the systems are 119.8164 MWh for pipeline A, 146.788 MWh for pipeline B, and 149.485 MWh for pipeline C.

3.5 Energy of Pump

At charging mode, the flow rate of water will decline at the time. It is also caused by the changing of hydrostatic pressure.

$$\frac{dZ_H}{dt} = \frac{v.A_p}{\left(A_H(Z_H) + Z_H\frac{dA_H}{dZ_H}\right)} \qquad (13)$$

$$\frac{dZ_H}{dt} = \frac{12{,}56\sqrt{\dfrac{2g(Z_H(t) - Z_L(t))}{\left(K + \dfrac{0{,}184}{\left(\frac{\rho v D}{\mu}\right)^{0.2}}\dfrac{L}{D}\right)}}}{(18849.8\,Z_H(t) + 531222)} \qquad (14)$$

Equation (13) is applied to calculate the increasing rate of water level in the lower reservoir. Substituting variables of the study site into (13) gives (14).

$$\frac{dZ_L}{dt} = \frac{-v.A_p}{\left(A_L(Z_L) + Z_L\dfrac{dA_L}{dZ_L}\right)} \qquad (15)$$

$$\frac{dZ_L}{dt} = \frac{-12{,}56\sqrt{\dfrac{2g(Z_H(t) - Z_L(t))}{\left(K + \dfrac{0{,}184}{\left(\frac{\rho v D}{\mu}\right)^{0.2}}\dfrac{L}{D}\right)}}}{(44348\,Z_L(t) + 2000000)} \qquad (16)$$

Furthermore, (15) is applied to calculate the declining rate of water level in the lower reservoir. Substituting the study site into (15) gives (16).

Furthermore, the flow rate of the water in charging mode can be found by solving the root of the flow rate quadratic equation below.

$$\frac{K\rho}{2}v_{pump}^{\ 3} + \frac{0.184.L.\rho}{\left(\frac{\rho D}{\mu}\right)^{0.2}.2D}v_{pump}^{\ 2.8} + \rho g(Z_H - Z_L)v_p \qquad (17)$$
$$\eta\frac{W_{pump}}{A_p} = 0$$

Where :

$_{pump}$: the flow rate of the pumping system (m/s)

η : efficiency of the pumping system

To determine the value of Z_H and Z_L, Runge-Kutta numerical method has to be applied with the certain initial condition. The initial condition of charging mode is equal to the final condition of discharging mode. So, the initial condition of the design A is $Z_H(0)$ = -9.930 m and $Z_L(0)$ = -43.685 m, the initial condition of the design B is $Z_H(0)$ = -26.080 m and $Z_L(0)$ = -40.734 m, and the initial condition of the design C is $Z_H(0)$ = -20.313 m and $Z_L(0)$ = -41.325 m.

The charging mode would be operated until the water level is about 0 m. The time needed to pump the water from lower to the upper reservoir in each pipeline is 5.75 hours for pipeline A, 12.25 hours for pipeline B, and 10 hours for pipeline C.

$$E_{pump} = W_{pump} \times t \qquad (18)$$

Where :

P_{ump} : energy needed by pump (MWh)

We_{pump} : power of pump (MW)

t : time (h)

Then, the amount of energy needed by pump can be solved by using (17)

Pump power needed by the PSH system is assumed to equal to the maximum power produced by the turbine. Based on the (17), the energy needed by the pump for the design A, B, and C is 158.7805 MWh, 261.17 MWh, and 233,69 MWh respectively.

3.6 Efficiency of The System

$$\eta_{system} = \frac{E_{turbin}}{E_{pompa}} \times 100\% \qquad (19)$$

The efficiency of the PSH system is obtained by using (18) above. The efficiency of the PSH system in each layout design of pipeline in this study are 75.460% for the design A, 56.204% for the design B, and 63.967% for the design C (see Table 1).

4 Results and Discussion

The starting operation time based on the natural water availability is 14.7 years from the year 2014. In that time, the upper reservoir water level is -0.087 m and the lower reservoir water level is -48.002 m. The efficiency of the PSH system is influenced by some variables. They are the pipeline design, volume of water that will flow, headloss and operational time per cycle.

Table 1. Comparison of the Three Designs

Length	Based on the layout design, the length of design A < design C < design B. The longer pipeline, the bigger headloss, and friction generated
Maximum Turbine Power	The maximum turbine power of design A < design C < design B. The difference is caused by the amount of head loss generated. The bigger headloss generated, the less maximum turbine power generated
Operational Cycle	The pump and turbine operational time of design A < design C < design B. The longer pipeline, the more time needed to flow the water. The operational time difference per cycle between pump and turbine is 0.25 hours for pipeline A and 0.5 hours for pipeline B. Pipeline C has the equal pump and turbine operational time per cycle.
Energy per Cycle	The energy generated per cycle of design C < design B < design A. The design A has the lowest energy generated than the others, although it has the smallest head loss and the highest maximum turbine power. It is caused by its flow rate, which is the smallest than others. The highest flow rate is in the design B, but its generated energy is lower than that generated by the design C. The higher head loss in the design B causes lower generated energy.
Efficiency	The efficiency generated by the system per cycle of the design A < the design C < the design B.

References

[1] S.H. Siahaan, T.A. Murwanto, S. Dolant, "Inovasi Teknologi PLTMH dan PLTB Lembaga Litbang Pemerintah". Jakarta, LIPI Press, 2011 (in Bahasa Indonesia)

[2] Badan Perijinan dan Penanaman Modal Daerah (BPPMD) Kalimantan Timur. "Penyusunan Kajian Investasi di Lahan Eks Tambang". Technical Document. Samarinda, 2012 (in Bahasa Indonesia)

[3] H.M. Ramos, M.P. Amaral, D.I.C. Covas, Journal of Water Resource and Protection **6**, 1099-1111 (2014)

[4] J. Olsen, K. Paasch, B. Lassen, C.T. Veje, Journal of Energy Storage **2**, 54-63 (2015)

[5] S. Kucukali, Renew. Sust. Energ. Rev **37**, 502 – 508. (2014)

[6] S. Padron, J.F. Medina, A. Rodriguez, Energy **36,** 6753-6762. (2011)

[7] D.A. Katsaprakakis, D.G. Christakis, I. Stefanakis, P.Spanos, N. Stefanakis, Energy **55**, 619-630 (2013)

[8] BMKG, "Jumlah Curah Hujan dan Jumlah Hari Hujan di Stasiun Pengamatan BMKG 2000-2013". Accessed from www.bps.go.id on April 1, 2016.

[9] S.M. Indriani, Analisis Kebutuhan Air, Irigasi Pangan dan Energi PLTMH (Pembangkit Listrik Tenaga Mikrohidro) Sungai Bayang, Kecamatan Bayang Utara, Kabupaten Pesisir Selatan. Thesis. Departement of Nuclear Engineering and Engineering Physics, Faculty of Engineering, Universitas Gadjah Mada, Yogyakarta, 2016. (in Bahasa Indonesia)

Anti Surge Control of Centrifugal Compressor at PT. Pertamina EP Asset 2 Field Pendopo

Muhammad L. Hakim[1], *Balza* Achmad[1,*], and *Juwari* P. Sutikno[2]

[1] Department of Nuclear Engineering and Engineering Physics, Faculty of Engineering, Universitas Gadjah Mada, Yogyakarta, Indonesia
[2] Department of Chemical Engineering, Sepuluh November Institute of Technology, Surabaya, Indonesia
[*]balzach@gmail.com (corresponding author)

Abstract. Gas Compression Package (GCP) is a vital system in the distribution of gas. Pertamina EP, in this regard as a state-owned company managing oil and gas on the upstream, is in the project phase of purchasing and installing compressor facilities. GCPs are used to increase gas pressure hence it can be transferred from gas wells to industrial areas. Dynamic analysis becomes important to observe the performance of the compressor and to ensure safety during operation, in order to avoid surge phenomenon. Surge is an undesirable phenomenon in the form of a backward flow of gas. The surge may damage the internal components of a centrifugal compressor, hence needs to be prevented from happening. The anti-surge controller manipulates anti-surge valve to give actions in the recycle lines when a surge occurred. PID controller is used for controlling and stabilizing the system. PID tuning is determined using Auto Tuning Variation method. This method is based on the process frequency response from relay oscillations. After the dynamic model of the plant had been developed, some simulations were carried out with various scenarios. The scenarios were based on a variety of issues that may occur in a gas compression system. The applied scenarios were a decrease in flow rate on the input side, a pressure drop in gas wells, and changes in gas composition. The control configuration of gas compression system was designed and modified in order to compensate disturbances and avoid surge phenomenon.
Keywords—anti-surge, centrifugal compressor, dynamic simulation, PID controller.

1 Introduction

Natural gas is one of the natural resources that have a broad range of utilization such as for cooking, for generating electricity in power plants, and as raw material for processing of fertilizer. South Sumatra, a province in Indonesia, has very abundant natural resources, including natural gas. According to the government of South Sumatra Province, natural gas reserves are found in Musi Banyuasin, Lahat, Musi Rawas, and Ogan Ogan Ilir, which reach 7238 Billion Standard Cubic Feet per Day (BSCFD). The production during the last 4 years on average was 2,247,124 Million Metric Standard Cubic Feet per Day (MMSCFD) [1]. PT. Pertamina EP is a state-owned company which engaged in the upstream oil and gas. Currently, PT. Pertamina EP produced approximately 120,000 Barrels of oil Oil per Day (BOPD) and around 1,003 MMSCFD of natural gas [2]. PT. Pertamina EP plans to increase its production capacity of 100 MMSCFD to 150 MMSFD by building a new compressor facility in Assets 2 Field Pendopo. This project is divided into 3 main section, i.e., gas separator system, gas compression system, and fuel gas-seal gas system. This paper focuses on the gas compression system which consists of three trains. Each train consists of unit operations such as medium pressure (MP) scrubber, centrifugal compressor, air cooler, high pressure (HP) scrubber, level control valve (LCV), and anti-surge valve (ASV).

A centrifugal compressor is the main part that is very vital for transporting natural gas from well to industrial, commercial, and residential areas. During transportation, the pressure of the natural gas will decrease because of the friction that happens along the wall of pipes, the pressure drop in unit operation, slug flow, etc. Another vital problem is called surge phenomenon because it disrupts the transport of natural gas. Surging is an undesired event which occurs in the form of reverse flow from the discharge of compressor to the suction of compressor in centrifugal compressors and axial compressors.

Surging is undesirable because it can cause many problems. If no action is taken, the surge will result in damage to the internal components of the compressor. It is not desirable for the company as the internal components have to be changed often while the costs of internal compressor parts are very expensive. However, appropriate precautions can be taken into account to prevent the surge. There are few symptoms of a surge, i.e., there will be a low system flow, increases of discharge temperature, extreme changes in current motor, as well as violent fluctuations in discharge pressure.

Prevention of surging can be done by installing either blowdown valve (BDV) or ASV in the recycle pipe.

https://doi.org/10.1051/e3sconf/20184201010

However, BDV causes too much energy loss, because gas is discarded and accommodated in the tank. It is better to use ASV in order to minimize the waste of energy. ASV has a set point above the minimum flow which is regulated by a controller namely anti-surge controller (ASC). ASC controls the opening of ASV when some of the symptoms of surging are about to occur. The way to find out when the occurrence of surging is it can be observed on the compressor curve which is plot based on the flow versus head. In the compressor, the curve is divided into several lines to observe surge those are surge control line (SCL), backup line (BL), and surge line (SL). Surging can be observed when the flow falls to the set point so that the compressor operating point on the curve crosses the SL.

2 Methodology

This research was conducted in several steps, i.e., developing dynamic models, analysis of the performance of PID controllers, and performing simulation scenario.

The initial simulation model was created in a steady state. The processes that occur in HYSYS simulation follow thermodynamic laws, mass balance, and energy balance. The dynamic simulation was then created by inputting data from Pertamina project into this initial model. In order to make a dynamic simulation, it needed sizing data of the unit operations and valves, the piping data, compressor performance, and the surge flow data which obtained from the vendor. The dynamic model that had been created should be stabilized by designing a PID controller.

There were some controlled parameters such as level, pressure, temperature, flow, and speed of the compressor in the plant. PID control parameters were determined using Auto Tuning Variation (ATV) method. A PID ATV method developed by Astrom-Hagglund is one of the auto-tuning technique that is simple and most appropriate to control process, which has been successfully applied to industries for over 15 years. This method is based on the response of process frequency of oscillation relays [3-6].

The last step was performing simulations with various scenarios. The scenarios were based on the problems that might occur in a gas compression system, i.e., a decrease in flow rate, a pressure drop in gas wells, and changes in input gas compositions. The scenarios can be analyzed in terms of operation and safety design modifications offered when compared to existing designs.

3 Results and Discussion

In order to determine the control parameter which is employed are based on the process conditions and design demands. PI controllers are commonly used in industries rather PID controllers because of many factors such as the tolerated offset, the presence of noise, excessive dead time, and small capacity.

After determining control parameters, the next step is tuning with Auto Tuning Variation (ATV) method, while for Anti-Surge Controller (ASC) tuning done by trial and error with certain procedures. Similar to Ziegler-Nichols method in PID tuning, ATV method employs K_{cu} and P_u. K_{cu} is controller gain that Products the limit cycle, while P_u is the period taken from limit cycle. The following calculations are to determine the ultimate period P_u, ultimate gain K_{cu}, controller gain K_c, and integral time T_i.

$$K_c = \frac{K_u}{3.2} \quad (1)$$

$$T_i = 2{,}2P_u \quad (2)$$

The control parameters followed the formulas as in Table 1. It is recommended for use Tyreus-Luyben formula for tuning that suitable for unit operation in Chemical process [7].

Table 1. Settings of PID controller based on the continuous cycle method [4]

Ziegler-Nichols	K_c	T_i	T_d
P	$0{,}5\ K_{cu}$	-	-
PI	$0{,}45K_{cu}$	$P_u/1{,}2$	-
PID	$0{,}6\ K_{cu}$	$P_u/2$	$P_u/8$
Tyreus-Luyben	K_c	T_i	T_d
PI	$0{,}31K_{cu}$	$2{,}2P_u$	-
PID	$0{,}45K_{cu}$	$2{,}2P_u$	$P_u/6{,}3$

For the algorithm of PID controller for the auto-tuning method, there is some additional parameter for PID controller algorithm, that is phase margin φ_m and design parameter α. The equation to determine the PID controller using ATV is as follows [7].

$$K_c = K_{cu}\cos(\varphi_m) \quad (3)$$

$$T_i = \alpha\, T_d \quad (4)$$

$$T_d = \frac{\tan(\varphi_m)\sqrt{(4/\alpha)+\tan(\varphi_m)}}{2\omega_c} \quad (5)$$

$$\omega_c = \frac{2\pi}{P_u} \quad (6)$$

Based on the formula and algorithm for PID tuning obtained the value of K_c, T_i, and T_d for all of the controller. There are total 24 controllers in the modified design which is eight controllers in each train. The value of K_c, T_i, and T_d are summarized in Table 2.

The dynamic model created in the process simulator and the scheme of gas compression systems are presented in Fig. 1 and Fig. 2.

Table 2. The results of tuning all control parameters

No	Label	Mode	SP Mode	SP	PV	OP	Controller	Kc	Ti	Td
1	LIC-2331	Auto	Sp Local	1000	1000.003	16.3094	PI	31.1	0.802	
2	LIC-2361	Auto	Sp Local	825	824.994	44.5636	PI	77.2	1.83	
3	LIC-2231	Auto	Sp Local	1000	1000.212	17.8985	PI	31.8	0.828	
4	LIC-2261	Auto	Sp Local	825	824.927	45.1033	PI	75.4	1.56	
5	LIC-2131	Auto	Sp Local	1000	999.987	16.6827	PI	22.7	0.512	
6	LIC-2161	Auto	Sp Local	825	824.991	43.9585	PI	75.6	1.55	
7	K-234: Surge Controller	Auto	Sp Local	67314.236	97621.723	0	PI	4	1.5	
8	K-224: Surge Controller	Auto	Sp Local	67330.445	99202.221	0	PI	4	1.5	
9	K-214: Surge Controller	Auto	Sp Local	67294.184	97911.148	0	PI	4	1.5	
10	TIC-2351	Auto	Sp Local	113	113.000	63.1747	PID	0.627	0.169	3.75×10^{-2}
11	PIC-1101	Auto	Sp Local	305	305.001	39.7273	PI	0.85	7.87×10^{-3}	
12	TIC-2251	Auto	Sp Local	113	112.892	60.651	PID	1.64	0.103	2.29×10^{-2}
13	PIC-0801	Auto	Sp Local	305	305.006	41.8596	PI	0.848	7.88×10^{-3}	
14	TIC-2151	Auto	Sp Local	113	113.001	62.9069	PID	0.955	0.119	2.63×10^{-2}
15	PIC-0501	Auto	Sp Local	305	305.001	41.0462	PI	0.849	5.63×10^{-3}	
16	PIC-2241	Auto	Sp Local	771	771.272	80.1118	PI	0.193	3.81×10^{-2}	
17	SIC-2241	Auto	Sp Remote	11455.994	11456.349	50.3773	PI	0.632	6.49×10^{-3}	
18	PIC-2341	Auto	Sp Local	771	771.000	79.6821	PI	0.197	5.65×10^{-2}	
19	SIC-2341	Auto	Sp Remote	11394.536	11394.536	49.557	PI	0.705	7.32×10^{-3}	
20	PIC-2141	Auto	Sp Local	771	771000	79.7459	PI	0.19	3.58×10^{-2}	
21	SIC-2141	Auto	Sp Remote	11403.663	11403.663	49.7234	PI	0.679	6.30×10^{-3}	
22	TIC-2251-1	Auto	Sp Local	93.25	93.309	57.122	PID	3.07	1.39×10^{-1}	3.10×10^{-2}
23	TIC-2351-1	Auto	Sp Local	93.25	93.250	57.5831	PID	3.09	1.39×10^{-1}	3.09×10^{-2}
24	TIC-2151-1	Auto	Sp Local	93.25	92.880	53.8588	PID	3.07	1.39×10^{-1}	3.10×10^{-2}

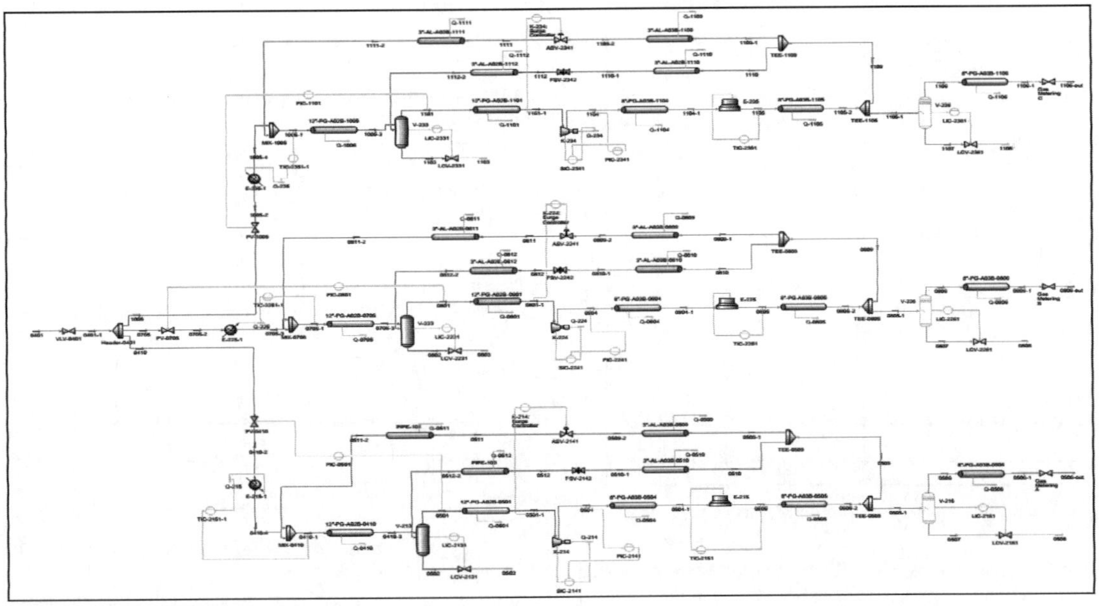

Fig. 1. The dynamic model of three train gas compression system

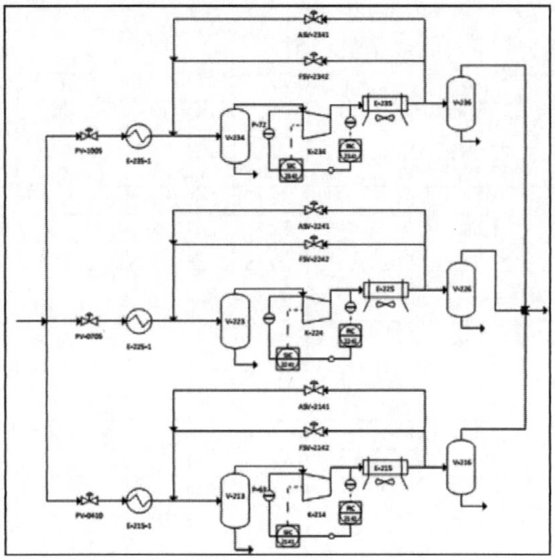

Fig. 2. Scheme of three train gas compression system

3.1 Scenario 1 : modified design

This first scenario demonstrated the importance of modified design in creating the model using HYSYS [8-11].

1) *Case 1: By using load sharing, the flow rate was decreased by 62% from 153.3 MMSCFD to 58.28 MMSCFD*

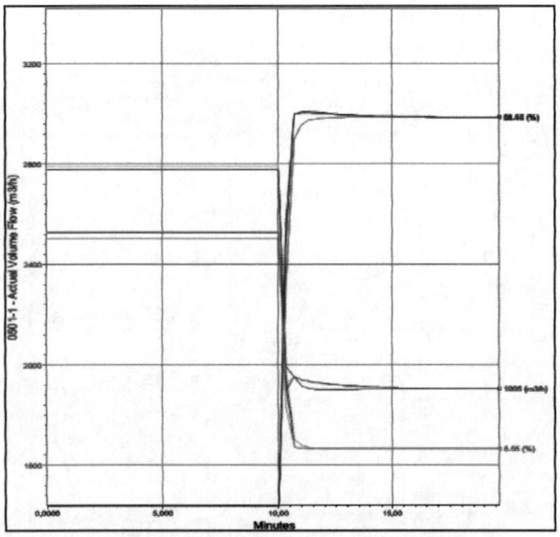

Fig. 3. The Result of load sharing for three train compressors

Fig. 3 shows the result of the simulation with three load sharing. When the flow rate dropped to a certain value, it caused surging. The above simulation conducted for 20 minutes by decreasing the flow rate from 153.3 MMSCFD to 58.28 MMSCFD. When surging was about to happen, ASV 2141, ASV-2241, ASV 2341 aggressively opened at 58.94%, 59.27%, 55.32% to prevent surging. The purpose of the installation of gas compression systems in parallel was to balance the load. When there was a surging, the pressure control valve (PV) would be split into three train loads with equal load with current openings steady at 8.5%.

2) *Case 2: The flow rate of the working fluid was reduced from 153.2 MMSCFD to be 65.71 MMSCFD, or reduced by 57.108% and the valve on the discharge failed.*

Table 3 and Table 4 show the comparison between the system which was equipped with control performance and load sharing, and the system without them. Compressor discharge pressure increased when the flow decreased, and discharge valve failed. In Table 4, when the pressure rose because of discharge valve failed, it can be overcome by increasing the RPM of the compressor. In Table 4, pressure rise cannot be anticipated, thus exceeding the design pressure of the unit operation in the discharge. This will certainly endanger the operating unit.

Table 3. The pressure of each train without control performance and load sharing

Parameter	K-214	K-224	K-234
Compressor Suction Pressure (psig)	317.4	317.5	317.6
Max Pressure. (psig)	320.3	320.3	320.3
Compressor Discharge Pressure *(psig)*	967.9	971	968.3
Min Discharge Pressure (psig)	750.0	750.0	750.0
Air Cooler Operating Pressure (psig)	770.0	770.0	770.0
Air Cooler Design Pressure (psig)	950.0	950.0	950.0
HP Scrubber OPerating Pressure (psig)	760.0	760.0	760.0
HP Scrubber Design Pressure (psig)	950.0	950.0	950.0

Table 4. The pressure of each train to control the performance and load sharing

Parameter	K-214	K-224	K-234
Compressor Suction Pressure (psig)	304.9	304.9	304.9
Max Pressure. (psig)	320.3	320.3	320.3
Compressor Discharge Pressure *(psig)*	771.0	771.0	771.0
Min Discharge Pressure (psig)	750.0	750.0	750.0
Air Cooler Operating Pressure (psig)	770.0	770.0	770.0
Air Cooler Design Pressure (psig)	950.0	950.0	950.0
HP Scrubber OPerating Pressure (psig)	760.0	760.0	760.0
HP Scrubber Design Pressure (psig)	950.0	950.0	950.0

3) *Case 3: Comparison of design in the case of denser gas in the train K-224*

Table 5 provides the comparison of the two systems when the molecular weight of gas is varied. It can be inferred; the modified design is more efficient compared to the existing design due to pressure ratio and the power needed to operate each compressor.

Table 5. The pressure of each train to control the performance and load sharing

Parameter	Existing Design	Modified Design
Adiabatic Efficiency (%)	70.69	71.388
Polytropic Efficiency (%)	73.147	73.799
Power (kW)	2,345.19	2,256.99
Suction Pressure (psig)	307.2	304.9
Discharge Pressure (psig)	772.5	771
Pressure Ratio	2.446	2.459

3.2 Scenario 2 : Low Suction Pressure

On the second scenario, it is observed how the system responds to disturbances in the form of low suction pressure. Any disturbance can cause a surge, so the role of the Anti-Surge Controller (ASC) and Anti-Surge Valve (ASV) is important to overcome surge. ASC has parameters named surge control parameter (A, B, C, D) which must be calculated by the user. The equation of surge control parameters is mathematically written as follows.

$$h_m = A + B \times F_s + C \times F_s^2 + D \times F_s^3 \quad (7)$$

Where F_s (m^3/s) is the surge flow, h_m (m) is the head, and A (m), B ($m/m^3/s$), C ($m/m^6/s^2$), and D ($m/m^9/s^3$) are surge control parameters set by the user. Table 6 provides the result of surge control parameter.

Table 6. Surge Control Parameters

No	Parameter	Value
1	Parameter A (m)	-28,134.9
2	Parameter B ($m/m^3/s$)	113,450
3	Parameter C ($m/m^6/s^2$)	-57,205.7
4	Parameter D ($m/m^9/s^3$)	0
5	*Control Line* (%)	10
6	*Backup Line* (%)	8
7	*Quick Opening* (%)	40

It can be observed from Fig. 4 and Fig. 5 when a low suction pressure is occurring, the flow rate will decrease temporarily. At a certain value of the decreasing of flow rate, it will cause surge [9, 10]. It can be observed that the blue line (actual volume flow) approaches the green line (the set point of surge controller). In order to prevent surge, ASV-2141, ASV-2241, and ASV-2341 aggressively opened at 76,6%, 55,22 %, 65,63%.

There is a limit of decreasing of low suction pressure. The safety factor is one of the considerations that commonly used to limit the margin of operation [12,13]. Generally, the safety factor is set to 5% to 10%. The design temperature for unit operation air cooler and HP scrubber is 300 °F. Table 7 presents the safety factor for operating temperature for the system. In order to meet the demand of safety factor, discharge pressure that allowed is set 260-281 psig, and the suction pressure is set 246.1 to 267.8 psig hence that the temperature generated in the discharge of 270-285 °F.

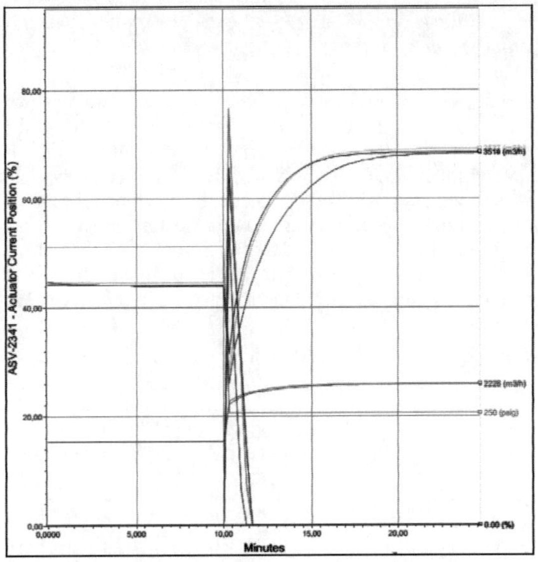

Fig. 4. The result of low suction pressure simulation

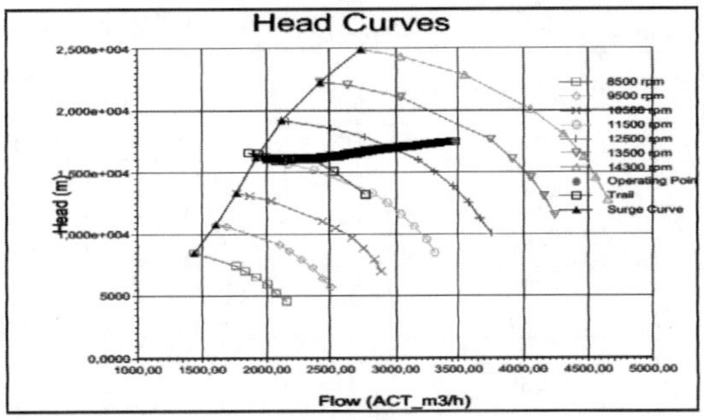

Fig. 5. Compressor performance for the low suction pressure

Table 7. Safety factor for The pressure of each train to control the performance and load sharing

10% Safety Factor for Temperature			
	Inlet	*Suction*	**Discharge**
Temperature (F)	101.2	93.12	270
Tekanan (psig)	281	267.8	771
Molar Flow (MMSCFD)	151	49.93	49.93
5% Safety Factor for Temperature			
	Inlet	*Suction*	**Discharge**
Temperatur (F)	101.2	93.04	285
Tekanan (psig)	260	246.1	771
Molar Flow (MMSCFD)	149.5	49.27	49.26

3.3 Scenario 3 : Change in Composition of Fluid

In Scenario 3, the simulation was conducted in order to observe the effect of a change in the composition of the fluid. Gas with heavier molecular weight named denser gas (+2 g/mol), otherwise gas with lighter molecular weight named leaner gas (-2 g/mol).

 https://doi.org/10.1051/e3sconf/20184201010

1) Case 1: leaner Gas (15,92 kg/m^3) and low molecular weight (17,06 g/mol)

Leaner gas contains very high methane about 95.496% and slight water content as shown in Table 8. Fig. 6 shows the performance of the compressor when the composition is changed to -2 g/mol. Gas with the lighter molecular weight will be easier to push. Thus, it does not need much discharge pressure. It has implications for the design demands; the pressure generated becomes less. Therefore, the RPM of the compressor will be increased by a speed controller, so greater flow rate occurs, then the pressure generated will reach the set point.

Table 8. Gas well composition with a molecular weight of 17.06 g/mol

	Mol fraction (mol %)	Gas phase (mol %)	Liquid phase (mol %)	Water (mol %)
Methane	77.771%	78.067%	8.413%	0.000%
Ethane	9.291%	9.319%	4.498%	0.000%
Propane	4.096%	4.100%	6.050%	0.000%
i-Butane	0.549%	0.548%	1.819%	0.000%
n-Butane	0.649%	0.646%	2.899%	0.000%
i-pentane	0.216%	0.213%	2.143%	0.000%
n-pentane	0.167%	0.163%	2.103%	0.000%
n-hexane	0.200%	0.187%	6.694%	0.000%
n-Heptane	0.383%	0.320%	30.689%	0.000%
n-Octane	0.083%	0.054%	13.850%	0.000%
n-Nonane	0.050%	0.021%	13.749%	0.000%
n-Decane	0.017%	0.004%	6.131%	0.000%
n-C11	0.000%	0.000%	0.000%	0.000%
H2S	0.000%	0.000%	0.000%	0.000%
H2O	0.516%	0.325%	0.089%	99.976%
Nitrogen	2.897%	2.909%	0.109%	0.001%
CO2	3.114%	3.125%	0.765%	0.023%

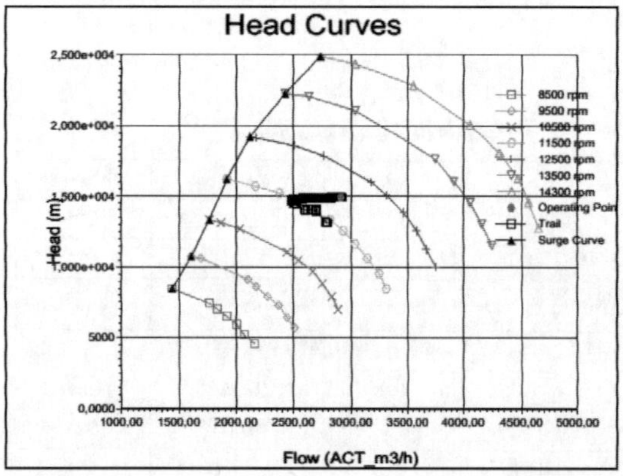

Fig. 6. Compressor Performance with a change in composition -2 g/mol

2) Case 2: Denser gas (15,92 kg/m^3) and high molecular weight (21,06 g/mol)

Otherwise, the denser gas contains methane, which is smaller than normal around 77.771% and more water content than normal as shown in Table 9. When the gas composition changed by increasing the molecular weight becomes +2 g/mol, it will have implications for the discharge pressure required more effort to depress the gas (Fig. 7). Thus the pressure would exceed the demands of the design. Therefore, the RPM of the compressor will be decreased by speed controller causes the flow rate become small so that the pressure generated in will reach the set points.

Table 9. Gas well composition with a molecular weight of 21.06 g/mol

	Mol fraction (mol %)	Gas phase (mol %)	Liquid phase (mol %)	Water (mol %)
Methane	95.496%	95.496%	25.295%	0.000%
Ethane	1.883%	1.883%	2.207%	0.000%
Propane	0.830%	0.830%	2.979%	0.000%
i-Butane	0.111%	0.111%	0.909%	0.000%
n-Butane	0.132%	0.132%	1.441%	0.000%
i-pentane	0.044%	0.044%	1.090%	0.000%
n-pentane	0.034%	0.034%	1.071%	0.000%
n-hexane	0.040%	0.040%	3.605%	0.000%
n-Heptane	0.078%	0.078%	18.704%	0.000%
n-Octane	0.017%	0.017%	10.938%	0.000%
n-Nonane	0.010%	0.010%	16.970%	0.000%
n-Decane	0.003%	0.003%	14.288%	0.000%
n-C11	0.000%	0.000%	0.000%	0.000%
H2S	0.000%	0.000%	0.000%	0.000%
H2O	0.105%	0.105%	0.069%	99.985%
Nitrogen	0.587%	0.587%	0.056%	0.000%
CO2	0.631%	0.631%	0.380%	0.014%

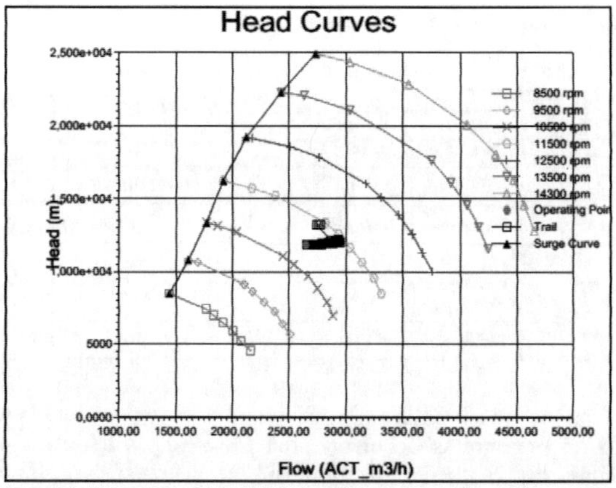

Fig. 7. Compressor Performance with a change in composition +2 g/mol

4 Conclusion

Simulation of gas compression systems in the dynamic state using a process simulator HYSYS has been successfully created and has stabilized. Models were made as detailed as possible in order to approach the actual condition of the plant. Some modifications were made such as the addition of control performance and load sharing. PID control parameters obtained by using Auto Tuning Variation method, almost the entire process variables are controlled by the PI controller except for temperature control using PID controllers. With scenarios such as a decrease in the input flow rate, low suction pressure, and changes in the composition will cause the compressor into the surge. In this Study, designed gas compression system has been able to prevent surge by regulating the ASC and ASV with the configuration offered on the simulation. Therefore the simulated disturbances are compensated by the system.

Acknowledgment

The authors would like to thank Mr. Dwi Yudha Prasetyatama and Pusat Studi Ilmu Teknik Universitas Gadjah Mada for providing data and consultations.

References

[1] BP3MD Sumatera Selatan, Mining and Energy 2014. Accessed on 13 April 2016. (in Bahasa Indonesia)

[2] Pertamina EP, Company Profile 2013. Accessed on 13 April 2016.

[3] Svrcek, W.Y. Mahoney, D.P. Young, R. Brent, *A Real-Time Approach to Process Control* (John Wiley & Sons, New Jersey, 2006)

[4] D. E. Seborg, T. F. Edgar, D.A. Mellichamp, *Process Dynamics and Control* (Wiley, New York, 1989)

[5] K.J. Astrom, T. Hagglund, *Automatic tuning of PID [controllers*
(Instrument Society of America, New Concord, 1988).

[6] C.C. Hang, K.J. Astrom, Q.G. Wang, J. Process. Contr **12**(1), 143-162 (2002)

[7] R. K. Herz, Closed-Loop Auto Relay Tuning - Auto Tune Variation, ATV. San Diego, University of California. 2012.

[8] C. Hansen. Dynamic Simulation of Compressor Control Systems. Esbjerg, 2008

[9] U.S. Paraguassu, Indian J. Sci. Technol **8** (2015).

[10] J. Liedman, R. Månsson, Dynamic simulation of a centrifugal compressor system, Gothenburg, 2013

[11] G.K. McMillan, *Centrifugal and Axial Compressor Control* (Instrument Society of America, New Concord, 1983)

[12] G. Towler, R. Sinnott, *Chemical Engineering Design: Principles, Practice and Economics of Plant and Process Design Second Edition* (Elsevier, Boston, 2013)

[13] R. Turton, R.C. Bailie, W.B. Whiting, J.A. Shaewitz, D. Bhattacharyya, *Analysis, Synthesis, and Design of Chemical Processes Fourth Edition* (Prentice-Hall, Upper Saddle River, 2012

E3S Web of Conferences **42**, 01011 (2018) https://doi.org/10.1051/e3sconf/20184201011
ASTECHNOVA 2016

The Study of System Planning of Assets Security on Marine Current Renewable Energy

Arinta Y. Wardani[1,*], *Juniarko* Prananda[1], *Irfan* S. Arief[1] ,and *Irwan* H. Suherman[2]

[1] Dept of Marine Engineering, Institut Teknologi Sepuluh Nopember, Surabaya, Indonesia
[2]Center of Research and Develpoment of Marine Geology, Ministry of Energy and Mineral Resources, Bandung, Indonesia

Abstract. Assets have a tendency to be the emergence of potential failures, with the exception of the components in the Marine Current Renewable Energy. The design review of it, related to indications of harm due to environmental conditions. Indications of the failure of the equipment are required to maintain the condition and quality. Problem Identification by Failure Mode Effect Analysis (FMEA) is needed to establish it's components which have a high potential failure. The potential failure of these components will be given as monitoring and specific control systems. Therefore, this research presents node components that need to be secured assets to provide security systems in critical condition with Risk Priority Number (RPN).

1 Introduction

Indonesia is an archipelago country which has many strait narrows. The condition brings potential effect. It will inflict the water flow speed increases become marine current. In addition, it is potential to generate new energy and renewable. Obviously, the circumstance could be an alternative to solve the electrical problem in this country.

Behind the potential condition, assets are an important thing in Marine Current Renewable Energy. They have a tendency to be the emergence of potential failure. It is necessary to do some preventive measures and analyzes the emergence of modes of potential failure and its consequences. The system planning of asset security design is based on international standard IEC 61400-25 (Wind turbines - Part 25: Communications for monitoring and control of wind power plants). These standards have proven successful applied to the wind power generation system; therefore it is able applied in Marine Current Renewable Energy.

2 Method

The research concerns in the scoring of failure criteria asset processed by FMEA (Failure Mode and Effect Analysis). FMEA (Failure Mode and Effect Analysis), a qualitative method to analyze the reliability and mitigate the occurrence of a failure or functional problems of an asset by identifying failure modes that may occur and their causes and consequences [4].

FMEA aims in the following [1]:

- As alternative design to the level of reliability and high safety, especially in the early step of the design of a system.

Ensure all failure modes and effects against an asset has been early predictable and anticipated.

Discover the list of potential failure and magnitude of the effects.

Give the basic qualitative analysis of reliability and availability.

Documenting the failure history of an asset that is useful for reference in the operation and maintenance in the future or even a reference for design changes if necessary.

As the basis for the development of priority corrective measures against assets.

Give justification for the determination of redundancy, failure detection system and automation.

In addition, FMEA is related to Risk Priority Number (RPN). RPN is the priority scale of the risk value for each component of the equipment [3]. RPN value obtained from multiplying the severity (S), occurrence (O) and detection (D). Severity is a value that represents the level of the consequences caused by the failure. The occurrence is a value that represents the frequency of occurrence of the failure mode. Then detection is a value that represents the level of difficulty in detecting the failure mode that appears. The more frequent mode of failure to appear, the higher the consequences and the more difficult level detection, the higher also value its RPN [2].

Table 1. Severity Ranking Criteria [4]

Rank	Description
1-2	Failure is of such minor nature that the customer (internal or external) will probably not detect the failure
3-5	*Failure will result in slight customer annoyance and/or slight deterioration of part or system performance*
6-7	*Failure will result in customer dissatisfaction and annoyance and/or deterioration of part or system performance*
8-9	*Failure will result in high degree of customer dissatisfaction and cause non-functionality of system*
10	*Failure will result in major customer dissatisfaction and cause nonsystem operation or non-compliance with government regulations*

Table 2. Occurrence Ranking Criteria [4]

Rank	Description
1	*An unlikely probability of occurrence during the item operating time interval. Unlikely is defined as a single failure mode (FM) probability < 0,001 of the overall probability of failure during the itam operating time interval*
2-3	*A remote probability of occurrence during the item operating time interval (i.e. once every two months). Remote is defined as a single FM probability > 0,001 but < 0,01 of overall probability of failure during the item operating time interval*
4-6	*An occasional probability of occurrence during the item operating time interval (i.e. once a month). Occasional is defined as a single FM probability > 0,01 but < 0,10 of the overall probability of failure during the item operating time interval*
7-9	*A moderate probability of occurrence during the item operating time interval (i.e. every two weeks). Probable is defined as a single FM probability > 0,1 but < 0,2 of the overall probability of failure during the item operating time interval*
10	*A high probability of occurrence during the item operating time interval (i.e. once a week). High probability is defined as a single FM probability > 0,20 of the overall probability of failure during the item operating time interval*

Table 3. Detection Ranking Criteria [4]

Rank	Description
1-2	*Very high probability that the defect will be detected. Verification and/or controls will almost certainly detect the existence of a deficiency or defect*
3-4	*High probability that the defect will be detected. Verification and/or controls will almost certainly detect the existence of a deficiency or defect*
5-7	*Moderate that the defect will be detected. Verification and/or controls will almost certainly detect the existence of a deficiency or defect*
8-9	*Low that the defect will be detected. Verification and/or controls will almost certainly detect the existence of a deficiency or defect*
10	*Very low that the defect will be detected. Verification and/or controls will almost certainly detect the existence of a deficiency or defect*

To compile an FMEA analysis is required to follow the existing standards, such as by making worksheets that follow the standard MIL-STD-1629A [5]. In case, there will be the number of failure modes that may emerge, thus, need a way to find where the failure mode that will be the priority to be solved. Hence, RPN (Risk Priority Number) is needed to determine the priority.

There are sequences in FMEA:

- Data Collecting, specification, system information and reviewing the system.
- Compile the functional block diagram (FBD)
  FBD used to describe the different components of the system interact each other. Thus, we can understand the critical path of a system failure.
- Analyze the failure mode and make the worksheets
  Determine potential failure modes as examples of the failure mode at the level of manufacturing, which is oversized, cracked, misassembled, and others. Then determine the potential effect of each failure mode. After that determines the potential causes of each failure mode. Then detect the failure through hardware, software, and others.
- Determine the *Risk Priority Number* (RPN)

RPN is a critical indicator of the failure mode in the system. It could be calculated by multiplying the severity (1-10), occurrence (1-10) and the detection rank (1-10) with the resulting rated on a scale of 1-1000. The scale has to mean, the smaller value of the RPN, the system is getting better. In the procedure, a system which has a high RPN, corrective actions would be performed on each failure mode.

- Forming the review team.
  The technical team has formed the purpose of assessing, comments and feedback on the worksheet that has been compiled and determine priorities based on RPN.
- Obtain the corrective action
- Obtain the monitoring system and control based on RPN.

There are components that must be clustered by FMEA analysis. The components divided into two categories, mechanical and electrical components. The components are included in the mechanical category are main shaft, turbines, bearings, shaft protective pipe, mooring system, and gear box. While the components are included in the electrical category are generators, batteries, inverters, and power control.

3 Result and Discussion

The results of this research aim to determine the value of Risk Probability Number (RPN) for each component. The scores of RPN are taken based on the analysis of FMEA tables that have been prepared. From those, can be obtained the average of RPN, which is shown in figure 1.

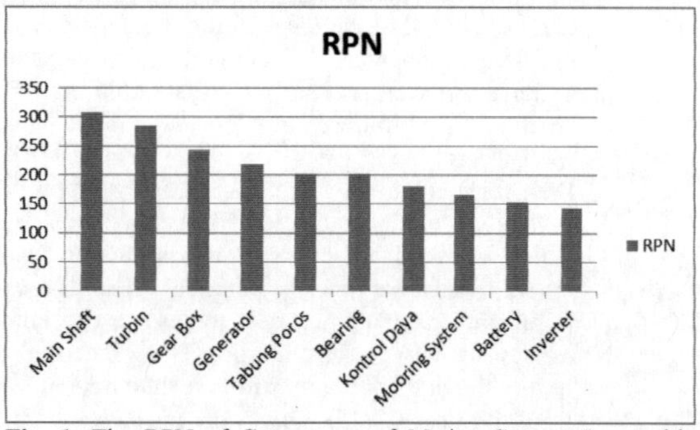

Fig. 1. The RPN of Component of Marine Current Renewable Energy

These are an analysis that shows the score of RPN (Risk Priority Number) of each component. For the main component (main shaft), there are four modes of failure (failure mode) with the potential to occur, which is broken by two different causes permanent changes shape, and not permanently deformed. So that, the failure mode of the four average values obtained RPN to the main shaft is equal to 308. Then, turbine components obtained three potential failure modes, namely blade broken in two different causes and deformed permanently accompanied by corrosion. Thus, from the three failure modes obtained average value RPN to Turbine is at 285.

Then there is a record for the mooring system, the component is not a critical component, even value its RPN was ranked third from the bottom. But, the value of its severity and occurrence are very lame. This shows that although the mooring system does not include a critical component of the system but the effect is very large if a failure occurs, meaning that the entire system should be switched off when mooring system failed. Therefore, it should be given special treatment (special treatment) of the mooring system by installing sensors that are considered capable of early detection of potential failures occur in the mooring system.

4 Conclusion

According to the research done, the following conclusion drawn:

1. The critical values of each component on Marine Current Renewable Energy have been shown in RPN, which scores are taken based on the FMEA analysis.
2. Based on RPN score, main shaft is critical component that needs to be secured in critical condition.
3. In the following research, FMEA analysis is the basis of the planning communications for monitoring and control system on Marine Current Renewable Energy.

References

1. A.G. Marbun, Failure Mode Implementation and Effect Analysis (FMEA) to Increase Reliability of Boiler Pipe, Jakarta: Teknik Industri, Universitas Mercubuana, (2013)
2. D. Isdarto, Risk Analysis of Operational Failure by Risk Failure Mode and Effect Analysis, Surabaya: Magister Manajemen Teknologi, Institut Teknologi Sepuluh Nopember (2014)
3. K.B. Artana, Risk Assesment Subsea Gas Pipeline, Surabaya: Guna Widya (2013)
4. M. Mraz, FMEA-FMECA (2005)
5. Tuncel and Alpan, Risk Assessment and Management for Supply Chain Networks : Case Study, Izmir: Dokuz Eylul University (2010)

 https://doi.org/10.1051/e3sconf/20184201012

Purification of Biodiesel Using Activated Carbon Produced from Cocoa Pod Husk

Devita Rachmat[1], *Aprillia D.* Agustin[1], and *Doty D.* Risanti[1]

[1]Department Engineering Physics, Institut Teknologi Sepuluh Nopember, Surabaya, Indonesia

Abstract. The purity level of waste cooking oil that use as raw material for biodiesel production has a strong influence on its fuel properties. In this study, activated carbon from cocoa pod husk activated with was used for purification in pre-treatment process of biodiesel production. The result showed that the activated carbon can decrease the water content and FFA and produce well-characterized biodiesel correspond to the Indonesian National Standard requirement.

1 Introduction

Biofuel has recently attracted huge attention in many countries because of its renewability, better gas emissions, and its biodegradability. Biodiesel is superior to conventional diesel in terms of its sulfur content, aromatic content and flash point. It is essentially sulfur free and non-aromatic while conventional diesel can contain up to 500 ppm SO_2 and 20-40% aromatic compounds and becomes one solution in reducing atmospheric pollution.

According to Utlu and Kocak [1], there was on average of a decrease of 14% for CO_2, 17.1% for CO and 22.5% for smoke density when using biodiesel.

Biodiesel can be produced from vegetable oils, however, the total manufacturing cost is unfavorable since the price could be 1.5 to 2 times higher than that for diesel [2]. On the other hand, the price of waste cooking oils (WCO) is 2-3 times cheaper than virgin vegetable oils which make WCO more attractive. The improper handling of unprocessed WCO promotes poor environmental impact due to its harmful disposal of these waste oils, such ends up in streams of the drains or leaches groundwater. It is known that the purity level of biodiesel from WCO has a strong influence on its fuel properties. Generally, biodiesel from WCO contains a large amount of water and a high proportion of free fatty acid (FFA). Thus, a pretreatment of WCO is needed to avoid saponification which leads to product separation and low fatty acid methyl ester (FAME). According to [3], pretreatment was successfully undertaken by using activated carbon and resulted in a lower water content and FFA.

There has been an increasing interest in the production of activated carbon from agricultural by-products and industrial waste. Cocoa pod husk [4-6] and different agricultural and forest wastes as raw materials, e.g. macadamia nutshell [7], rice husk [8-10], jatropha husk [11], wood apple shell [12], pine nutshell [13], oil palm shell [14], seed shell of the palm tree [15], sago waste [16], guava seed [17], almond shell, pistachio shell, walnut shell [18], olive mill waste [19] and others have been used in the production of activated carbon thereby adding value to these agricultural wastes and thus recycling them. This research focuses on the use of activated carbon prepared from cocoa pod husk using HCl as activating agents with a view to preparing, characterizing and investigating the adsorption effectiveness of the charcoal in the pretreatment of WCO.

2 Methods

Cacao pod wastes were dried in an oven at 100°C for 24 hours to remove the water content. They were subsequently crushed into powder and sieved using 60 mesh siever. The powder was then heated at 3 temperatures ranging from 250°C to 350°C for 2 hours. 37.5 gram of the powder was then mixed with 75 ml of HCl 2M, stirred for 4 hours at 80°C and kept at room temperature for 24 hours. The activated charcoal was subsequently heated again at 200°C for 3 hours. For phase present, surface area and functional groups of the activated charcoal were determined through XRD, FTIR spectroscopy, and SEM, respectively

The collected WCO was heated at 120°C for at least 5 minutes to remove the water content. After that, the activated charcoal was immersed in WCO for 10 hours and filtered to remove any solid impurities. The final product was then reheated at 120°C for 2 minutes.

Lye solution was made from a mixture of 156.25 ml methanol and 3.75 gram KOH under vigorous stirring for 5 minutes. For transesterification, the filtered WCO was reheated again at 120°C for 5 minutes and the temperature was gradually dropped to 55°C. Under stirring, 25 ml lye solution was added dropwise to the WCO for 1 hour. The result was kept for 8 hours until two layers were obtained, the upper was the methyl esters and the lower was the glycerol. Glycerol layer was discarded and the methyl ester layer was subsequently washed by using water and kept for 4 hours. The final solution was prefiltered again using whatman paper.

Properties of the purified biodiesel were measured according to the ASTM standards, i.e. ASTM D1298 for density, ASTM D445 for kinematic viscosity, ASTM D93-80 for flash point, ASTM D2500 for cloud point and ASTM D97 for pour point.

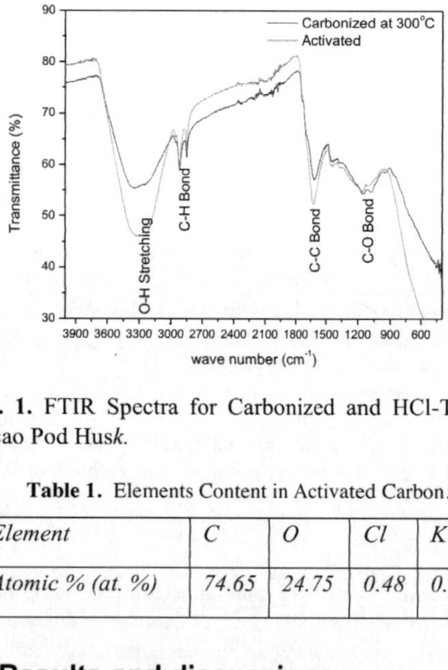

Fig. 1. FTIR Spectra for Carbonized and HCl-Treated Cacao Pod Hus*k*.

Table 1. Elements Content in Activated Carbon.

Element	*C*	*O*	*Cl*	*K*
Atomic % (at. %)	*74.65*	*24.75*	*0.48*	*0.12*

3 Results and discussions

3.1 Structure of Cacao Pod after carbonization

According to thermogravimetry results [20] the obtained weight losses during carbonization were 10.27%, 13.56 % and 17.64% at temperatures of 250, 300 and 400°C, respectively. The strong decrease in weight at 400°C corresponds to the decomposition of lignocellulose whereas at 250°C is attributed to the decomposition of hemicellulose [22]. Thermal degradation has taken place through heat supplied during the carbonization process [5].

Fig. 2. SEM Result of Activated Charcoal at Carbonized Temperature 300°C

3.2 Scanning Electron Microscope

Scanning electron microscopy (SEM) was employed for characterizing the surface morphology of the charcoals. Fig. 2 shows the SEM micrograph of the activated sample. The surface of the sample is rough and pores are not evenly distributed. This image can be analyzed that the carbonization and HCl treated only create gradual porosity. Table 1 illustrates the elemental analysis of the activated charcoal. The major component is carbon with some oxygen. The carbon majority is in agreement with the XRD results [20], while the existence of oxygen is in agreement with the FTIR analysis for the presence of cellulose, hemicellulose, and lignin.

3.3 Characterization of Biodiesel

The use of activated charcoal in the pretreatment of waste cooking oil influenced the characterization of the biodiesel. The characterization result shows in Table 2. which the values correspond to the Indonesian National Standard requirement.

The free fatty acid value should not be more than 1% because it may lead to the soap formation and hence lower the yield [21]. Improper handling of oil would cause it to oxidize in the presence of water to free fatty acid due to hydrolysis and oxidation of the triglycerides. Therefore only the charcoal activated at 300°C is suitable.

Table 2. Properties of Biodiesel Purified Using Activated Charcoal.

Property	Activated charcoal at carbonized temperatures			Indonesian National Standard
	250°C	300°C	350°C	
Density kg/m^3	869.5	872.5	865.5	850-890
KV @ 40C mm^2/sec	4.177	4.149	3.916	2.3-6.0
% FFA	1.28	0.43	1.024	
FP (°C)	190	175	180	Min. 100
CP (°C)	10.9	8.3	9	Max. 18
PF (°C)	17	13	14.1	Max. 28

4 Conclusion

The use of cocoa pod activated charcoal at 300°C in the pretreatment of waste cooking oil was shown to be successful in lowering the free fatty acid and able to meet the Indonesian National Standard requirement.

The authors would like to thank Directorate General of Higher Education Ministry of Research, Technology and Higher Education for the funding of this research through the Student Creativity 2016 Republic of Indonesia.

References

1. M.S. Kocak, E. Ileri, Z. Utlu, *Experimental study of emission parameters of biodiesel fuels obtained from canola, hazelnut, and waste cooking oils*, Energy Fuels, **21**, 3622-3626 (2007)
2. M. Ahmad, M.A. Khan, M. Zafar, S. Sultana, *Practical Handbook on Biodiesel Production and Properties*, CRC Press, Boca Raton FL USA (2013)
3. R.S. Putra, T.S. Juliantoa, P. Hartono, R.D. Puspitasaria, A. Kurniawan, *Pretreatment of Used-Cooking Oil as Feed Stocks of Biodiesel Production by Using Activated Carbon and Clay Minerals*, International Journal of Renewable Energy Development, **3(1)**, 33-35 (2014)
4. C. Theivarasu, S. Mylsamy, *Equilibrium and Kinetic adsorption studies of Rhodamine-B from aqueous solutions using cocoa (Theobroma cacao) shell as a new adsorbent*, IJEST, **2**, 6284-6292 (2010)
5. O. Bello, M. Ahmad, T. Siang, *Utilization of Cocoa Pod Husk for the removal of Remazol Black B reactive dye from aqueous solutions: kinetic, equilibrium and thermodynamic studies*, Trends in Appl Sci Research, **6**, 794-812 (2011)
6. F. Ahmad, W. Daud, M. Ahmad, R. Radzi, *Using cocoa (Theobroma cacao) shell-based activated carbon to remove 4-nitrophenol from aqueous solution: Kinetics and equilibrium studies*, Chem. Eng. Journal, **178**, 461– 467 (2011)
7. A. Ahmadpour, *The preparation of activated carbon from macadamia nutshell by chemical activation*, Carbon, **35**, 1723-1732 (1997)
8. F. Adekola, H. Adegoke, *Adsorption of blue-dye on activated carbons produced from rice husk, coconut shell and coconut coirpith*, Ife J. Sci., **7**, 151-157 (2005)
9. A. Jameel, A. Hussain, *Removal of heavy metals from wastewater using activated rice husk carbon as adsorbent*, Indian J. Environ. Protect, **29**, 263-265 (2009)
10. N. Khan, M. Faizal, I. Abustan, M. Azmier, *Effect of preparation conditions of activated carbon prepared from rice husk by $ZnCl_2$ activation for removal of Cu (II) from aqueous solution*, Int. J. Eng. Tech., **10**, 27-31 (2010)
11. K. Ramakrishnan, C. Namasivayam, *Development and characteristics of activated carbons from jatropha husk, an agro industrial solid waste, by chemical activation methods*, J. Environ. Eng. Manag., **19**, 173-178 (2009)
12. R. Malarvizhi, N, Sulochana, *Sorption isotherm and kinetic studies of methylene blue uptake onto activated carbon prepared from wood apple shell*, J. Environ. Protect Sci., **2**, 40-46 (2008)
13. T.A. Bal'burova, V.G. Shiretorova, E.V. Zoltoev, G.I. Khanturgaeva, *Production of activated carbons from Pine Nutshells*, Russ. J. Appl. Chem., **81**, 162-164 (2008)
14. C.L. Aik, J. Qipeng, *Adsorption of phenol by oil-palm-shell activated carbons*, Adsorption, **13**, 129-137 (2007)
15. S.Y. Gueu, A.G. Adouby, *Heavy metals removal in aqueous solution by activated carbon prepared from coconut shell and seed shell of the palm tree*, J. Appl. Sci., **6**, 2789-2793 (2006)
16. K. Kadirvelu, M. Kavipriya, C. Karthika, N. Vennilamani, S. Pattabhi, *Mercury (II) adsorption by activated carbon made from sago waste*, Carbon, **42**, 745-752 (2004)
17. I.A. Rahman, B. Saad, *Utilisation of Guava seeds as a source of activated carbon for removal of Methylene blue from aqueous solution*, Malay. J. Chem., **5**, 8-14 (2003)

18. J. Hayashi, T. Horikawa, I. Takeda, K. Muroyama, F. NasirAni, *Preparing activated carbon from various nutshells by chemical activation with K_2CO_3*, Carbon, **40**, 2381-2386 (2002)

19. C. Moreno, F. Carrasco, M.V. Lopez, M.A. Alvarez, *Chemical and physical activation of olive-mill waste water to produce activated carbons*, Carbon, **39**, 1415-1420 (2001)

20. D. Rachmat, A.D. Agustin, D.D. Risanti, *Cocoa (Theobroma cacao) pod-based activated carbon by acid activation*, 2nd International Seminar on Science and Technology (submitted) (2016)

21. Y.C. Sharma, B. Singh, S.N. Upadhyay, *Advancements in development and characterization of biodiesel: a review*, Fuel, **87**, 2355-2373 (2008)

22. K. S. Chuna, S. Husseinsyaha, H. Osmana, *Development of Biocomposites from Cocoa Pod Husk and Polypropylene: Effect of Filler Content and 3-Aminopropyltriethoxylsilane*, Polymers from Renewable Resources, **Vol. 5**, No. 4: 139-156 (2014)

 https://doi.org/10.1051/e3sconf/20184201013

Numerical study of the effect of geometry variation on the performance of innovative design wind speed enhancer

Theodorus T. Wibowo[1], *Faizal* H. Daulay, *Kutut* Suryopratomo, and *Rachmawan* Budiarto

[1]Department of Nuclear Engineering and Engineering Physics, Universitas Gadjah Mada, Yogyakarta, Indonesia

Abstract. Low wind speeds is one of the challenges of wind energy in several countries. The innovative design of wind speed enhancer is one of many solution of those challenges. The wind speed enhancer using the principle of constriction where the speed will be increased at a narrower area. The innovative design wind speed enhancer is unique because of the wind turbines are installed close to the ground surface. The uniqueness can reduce the cost of installation, maintenance, and can reduce the possibility of damage caused by lightning. The proposed of wind speed enhancer design was shown to be successful in increasing the wind speed at venturi that reach 2.26 faster than wind speed at environment for variation of the funnel gap and 2.35 faster than wind speed at environment for variation of the throat diameter. This result is better than existing design. The results of this study is 21.5% better than existing design for variation of funnel gap and 24.2% better than existing design for variation of the throat diameter.

1 Introduction

1.1. Background

In this electronic era, the electricity needs of the people of Indonesia is increasing, but there are still people who have not been reached by electric energy. In 2015 Indonesian government made a program called 35,000 MW. In that program, there is a use of fossil energy, and renewable energy.

Fossil fuels are all forms of energy derived from fossil both animals and plants. Fossil fuels have a long life cycle thus classified as non-renewable energy. Some examples of fossil fuels such as coal, oil and natural gas reserves are diminishing every year.

Renewable energy is the energy source that has a short life cycle. There are several examples of renewable energy such as solar energy (solar panels), wind energy (wind turbines), hydro energy, geothermal, and biomass. Renewable energy is one solution to meet energy needs in Indonesia. This article will discuss the solution of problems in the field of wind energy. The biggest problem of wind energy in Indonesia, almost all regions in Indonesia have low wind speed [12].

Wind speed in Indonesia is relatively low while wind turbines available on the market have the specifications for the high speed of wind [8], so it is necessary to increase the speed of the wind to harvest wind energy in Indonesia. By using the venturi principle, wind can be accelerated in a section where the wind turbines will be placed. The wind turbine will be placed close to the ground as it eases the 75process of installation, maintenance and to avoid damage due to lightning. A similar process has been performed by SheerWind with a project named INVELOX Wind Turbine in 2012-2014 [1-3]. In this research will be carried out numerical studies on several variations of geometry, namely: The variation of ratio between throat's diameter and the diameter of inlet venturi, and the variation of ratio between the gap of funnel and the length of funnel at Innovative design of wind speed enhancers. Numerical studies are conducted using ANSYS and outcomes to be achieved in this research is to obtain optimal geometry variations that increase the highest wind speed.

1.2. Literature

Many studies has been done to increase the capacity of wind turbines. Research conducted by Daryoush Allaei, and Yiannis Andreopoulos [1] is one of the research on wind energy issues. The research is to make the design of a device for increasing the speed of the wind, the device is called INVELOX. The device offers a concept to overcome the problem of wind energy in the regions with low wind speeds.

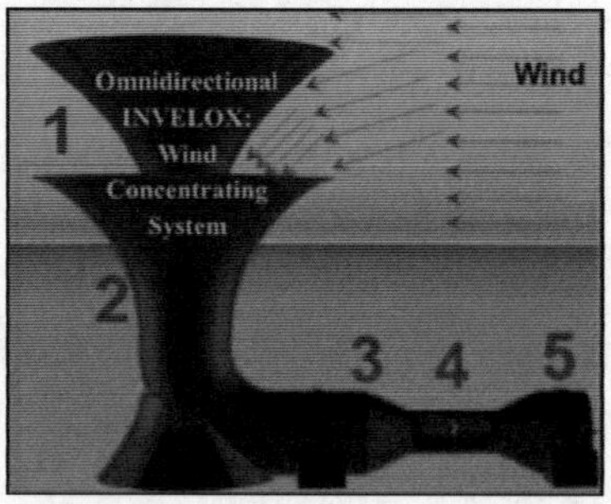

Fig. 1. Schematic INVELOX

INVELOX work scheme consists of 5 sections as shown in Figure 1. The first part is captured the wind from any direction via omnidirectional intake, then collected through a tunnel, after which the the wind is directed through the elbow to the venturi. The wind is accelerated on venturi and the wind turns the turbine so that electricity can be generated. The wind that had been through the wind turbines will be returned to the environment.

In the study conducted two methods: the method of CFD (Computational Fluid Dynamics) and experimental methods. CFD methods that have been carried out using the software ANSYS FLUENT and COMSOL Multiphysics, while the experimental method is done by taking data in the field in Chaska, Minnesota in 2012 and 2013.

Table 1. Data from CFD method

Model	Mesh size	Free stream [m/s]	*Venturi* velocity [m/s]		Speed ratio (SR)	
			Average	Max	Average	Max
ANSYS	Fine	6.71	10.6	12.1	1.58	1.80
COMSOL	Normal	6.71	10.6	12.1	1.58	1.80
	Fine	6.71	11.7	13.1	1.74	1.95

The experiment on CFD method using ANSYS and COMSOL resulted wind speed ratio reached 1.58 using ANSYS, and reached 1.74 in mesh size: fine using COMSOL.

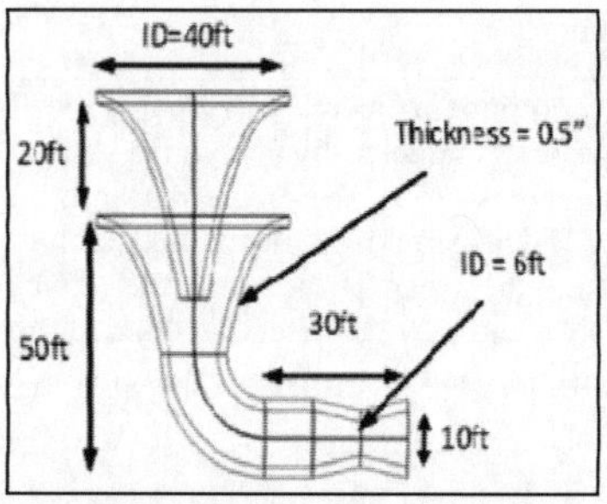

Fig. 2. INVELOX dimension

Figure 2 is a design experiment in Chaska, Minnesota. The results of this experimental method is a graphic of serial data displaying free stream wind, the wind speed at the venturi, and the speed ratio.

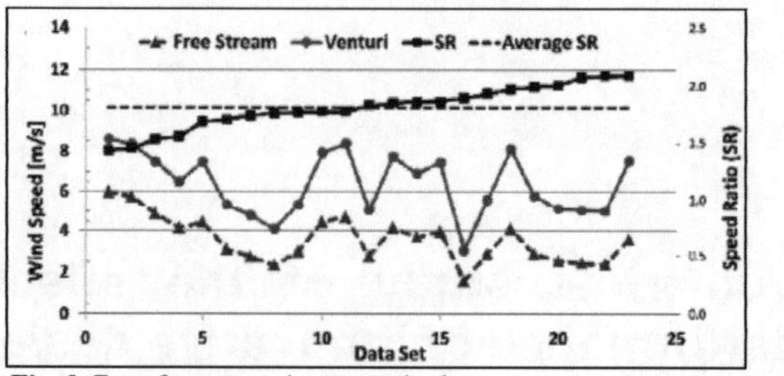

Fig. 3. Data from experiment method

Figure 3 was an experiment result by Daryoush Allei, and Yiannis Andreopoulos [1]. Based on research using CFD and experiment, INVELOX can increase the wind speed reaches the speed ratio 2. In the present study will use the base geometry similar to the geometry INVELOX and do variations on the intake funnel and a venturi section to improve performance enhancers wind speed

2 Research Methodology

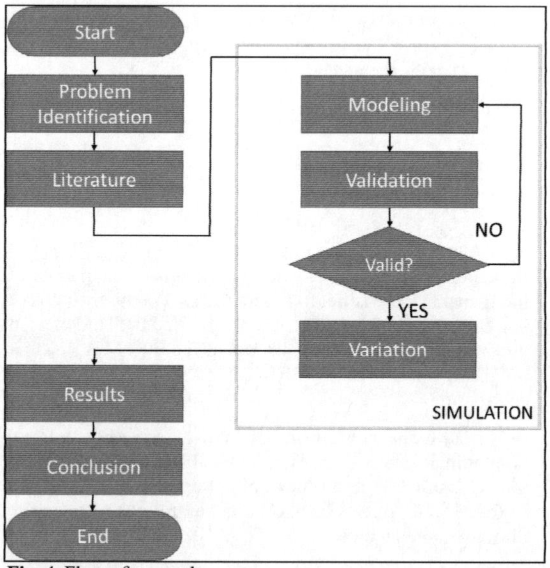

Fig. 4. Flow of research

Data used in this research is secondary data obtained from Daryoush Allaei and Yiannis Adreopoulos on paper "INVELOX: Description of a new concept in wind power and its performance evaluation". Then, the tools used in this research are shown in Table 2.

https://doi.org/10.1051/e3sconf/20184201013

Table 2. Research Tools

No.	Tools	Function
1.	ASUS A455L, *processor* Intel® Core™ i5-4210U CPU @2.4 GHz, *Installed memory* (RAM) 4.00 GB, 64-bit *operating system*	Hardware for simulation
2.	ANSYS 16.2	Software for design simulation

This study uses ANSYS software to perform numerical studies.

2.1. Modeling

Modeling is a process that is done in the research using an approach that is expected to resemble the real conditions in the field, the modeling will be done using Ansys.

2.1.1 Geometry and Boundary Conditions

The geometry of the wind speed enhancer has the same dimensions with the geometry INVELOX installed in Minnesota. Dimensions of geometry INVELOX contained in Figure 2 in Chapter I. besides geometry wind speed enhancer, virtual wind tunnel also needs to be made. This virtual wind tunnel created as the air domain. The geometry of the wind speed enhancer and virtual wind tunnel shown in the Figure 5.

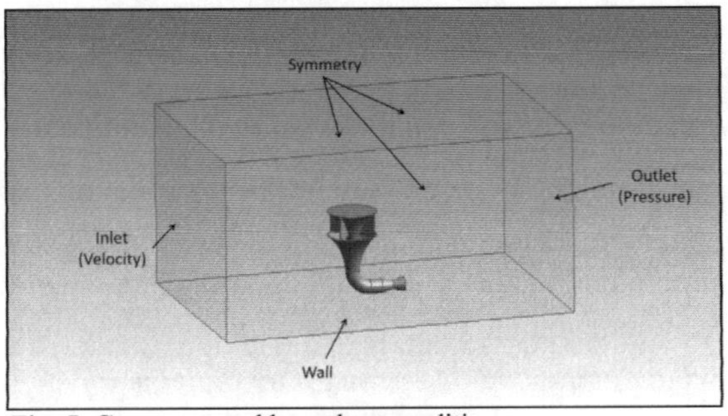

Fig. 5. Geometry and boundary condition

Virtual wind tunnel shaped like a beam with a length of 91.44 m (300 feet), width of 60.96 m (200 feet), and 45.72 m high (150 feet). Type of boundary conditions at the inlet is velocity and the outlet is the pressure. Turbulent Specifications methode at the inlet and outlet is intensity and length scale, turbulent intensity with a value of 5% and turbulent length scale of 1 m [1]. The vertical and the top side side of the beam set as wall with a slip condition and the bottom side of the beam set as wall with a no-slip condition because this side indicates the condition of the ground surface.

2.1.2 Numerical Settings

Numerical settings used for validation are as follows:

- Solver used in this study is the Pressure-based, with the segregated solver algorithms. Pressure-based solver is excellent in resolving cases at lower Mach number, and requires less memory [13].
- Model used in this study is the k-epsilon turbulence standard wall function.

2.1.3 Meshing

Meshing method used is tetrahedral meshing. Mesh is made finer at the venturi as the venturi section is an area of increasing wind speed, so it has a fairly high gradient. After meshing geometry shown in Figure 6.

Fig. 6. Meshing on wind speed enhancher

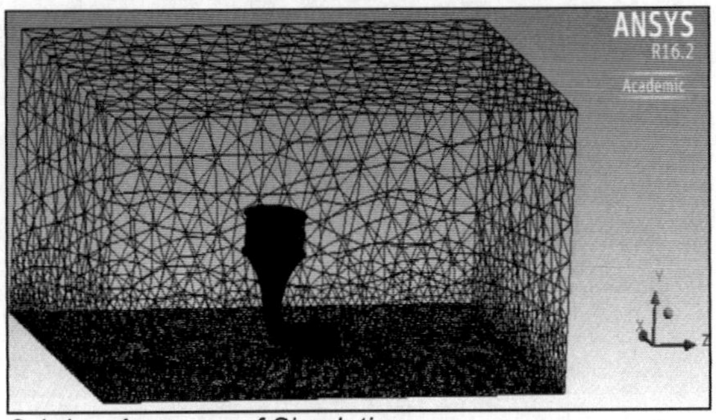

2.1.4 Accuracy of Simulation

The accuracy of the simulation results in this study were divided into 3 parts: quality mesh, numerical convergence, and mass imbalance.

1) Mesh Quality

The accuracy of the simulation results in this study were divided into 3 parts: quality mesh, numerical convergence, and mass imbalance.

Statistics	
Nodes	35296
Elements	173875
Mesh Metric	Skewness
Min	4.9559e-005
Max	0.96955
Average	0.24235
Standard Deviation	0.13001

Statistics	
Nodes	35296
Elements	173875
Mesh Metric	Orthogonal Quality
Min	0.12958
Max	0.99541
Average	0.85002
Standard Deviation	9.1547e-002

Fig. 7. Mesh metric quality: Skewness (left), Orthogonal (right)

In Figure 7, the average value of mesh skewness is 0.24 and the average value of orthogonal mesh quality is 0.85. Mesh can be good or not by comparing these values with the values shown in Figure 8.

Skewness mesh metrics spectrum

Excellent	Very good	Good	Acceptable	Bad	Unacceptable
0-0.25	0.25-0.50	0.50-0.80	0.80-0.94	0.95-0.97	0.98-1.00

Orthogonal Quality mesh metrics spectrum

Unacceptable	Bad	Acceptable	Good	Very good	Excellent
0-0.001	0.001-0.14	0.15-0.20	0.20-0.69	0.70-0.95	0.95-1.00

Fig. 8. Parameter of mesh quality [14].

From Figure 8, it can be seen that the quality of mesh this time are in the excellent category for skewness and very good for the orthogonal quality.

 https://doi.org/10.1051/e3sconf/20184201013

2) Numerical Convergence

Convergent numerically can be seen from the plot where the residual convergent condition is reached when the residual graph is straight or showed no decrease in error. In Figure 9, Convergent reached at iteration to 1000, but to ensure the error does not go down again, iterations performed until iteration 2500.

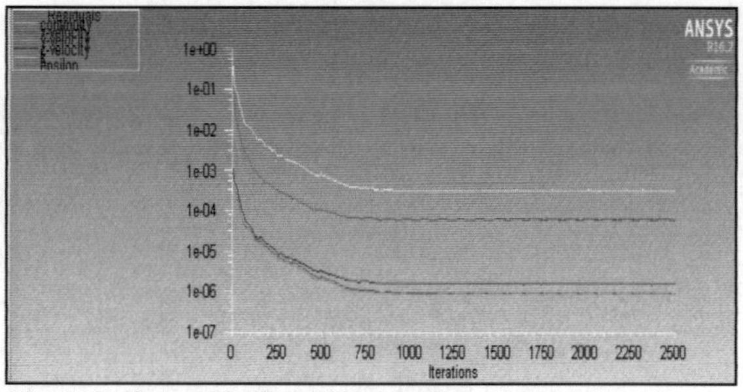

Fig. 9. Scaled residual graphic

3) Mass Imbalance

Mass balance can be demonstrated using the features of the report fluxes. Inlet and outlet is selected, then select compute as in Figure 10.

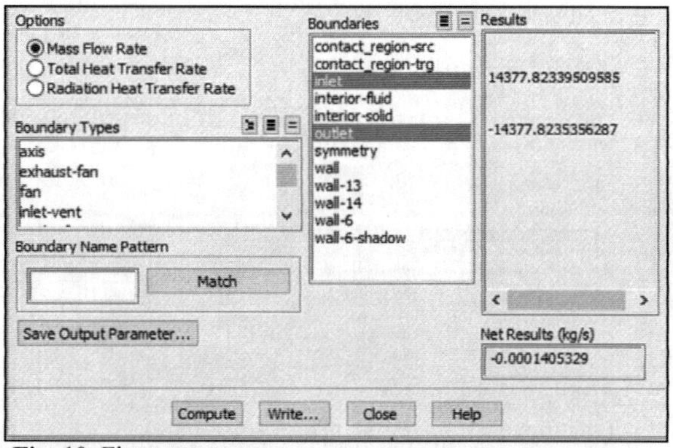

Fig. 10. Flux reports

Mass balance must be met in order to obtain satisfactory results, as shown by the error-value below 1% [13]. From figure 5.6, the value of net results is 0.0001 kg / s. This value is derived from the difference between the rate of mass in the inlet and outlet. The value of the net result shows the error is far below 1% and the mass balance is met.

2.2 Validation

This stage is the process of proving whether the simulation results using k-epsilon turbulence models have a value close to the experimental data. The experimental data are secondary data refers to the paper, INVELOX: Description of a new concept in wind power and its performance evaluation by Daryoush Allaei and Yiannis Adreopoulos.

Fig. 11. Validation of modeling with data experiment

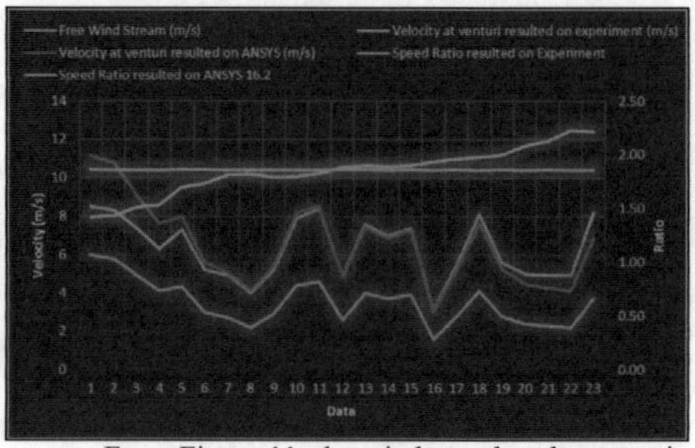

From Figure 11, the wind speed at the venturi from experimental data (orange line) and the data of simulation (dark blue line) shows the same pattern. wind speed ratio of the experiment (gray line) tends to rise while the simulation results (green line) is relatively constant. The difference between the data of simulation and experimental data is expected due to problem definition. Simulations carried out by the direction of the wind has not changed (one way) while the direction of the wind coming in the field can change. The difference from the simulation data and experimental data is not so significant that the model used is said to be valid. After validation, the next is to vary the geometry using models that have been valid.

2.3. Geometry Variation

2.3.1. Variation of the Funnel Gap

There are two variables that will be the reference to the variation of the length of funnel shown in Figure 12, namely:

- X = the length of the lower funnel.
- Y= the distance between the lower end of the lower funnel to the lower end of the upper funnel.

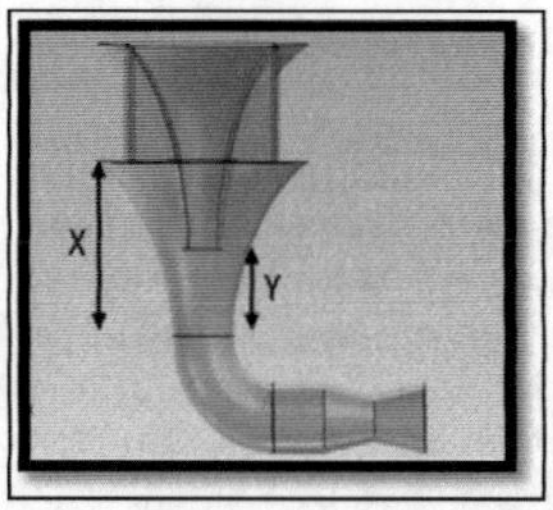

Fig. 12. Funnel gap variation

The length of the lower funnel maintained at 30 feet, according to the problem definition. In this study, free stream velocity, and the distance between the lower end of the lower funnel to the lower end of the upper funnel will be varied as in the following description.

X = 30 feet
Y = 0 feet, 7,5 feet, 15 feet, 22,5 feet, 30 feet

https://doi.org/10.1051/e3sconf/20184201013

FWS = 4 m/s, 3 m/s, 2 m/s, 1 m/s

Variation of geometry is simply the ratio of y to x (y/x) where the ratio of the length of funnel is 0, 0.25, 0.50, 0.75, and 1 can be seen in Figure 13.

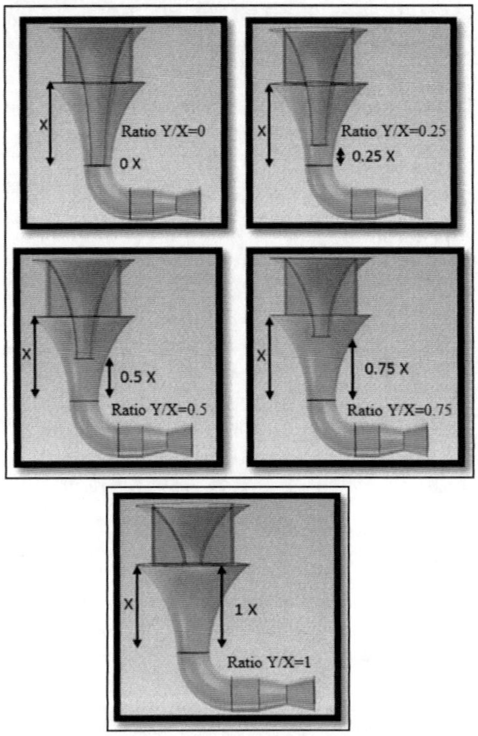

Fig. 13. Ratio of funnel gap variation

2.3.2. Variation of Rratio Throat's Diameter

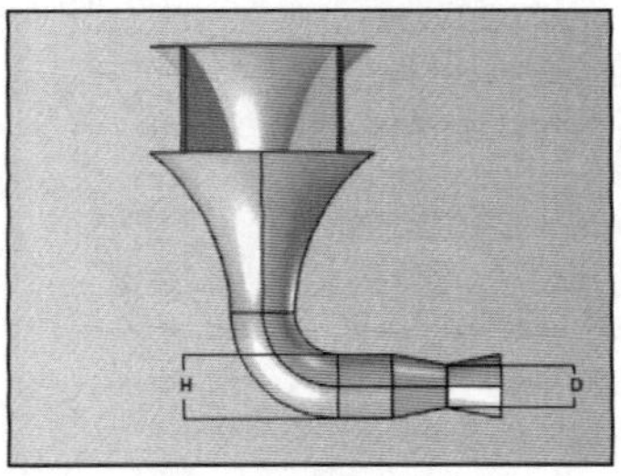

Fig. 14. Ratio of funnel gap variation

There are two variables that will be the reference to the variation of the length of funnel shown in Figure 14, namely:

- H = Diameter of inlet venturi
- D = Diameter of throat

FWS (m/s) = 1, 2, 3, 4

H = 10 feet

D = 9 feet, 8 feet, 7 feet, 6 feet, 5 feet,4 feet, 3 feet, 2 feet

Ratio = 1.11, 1.25, 1.43, 1.67, 2.00, 2.50, 3.33, 5.00

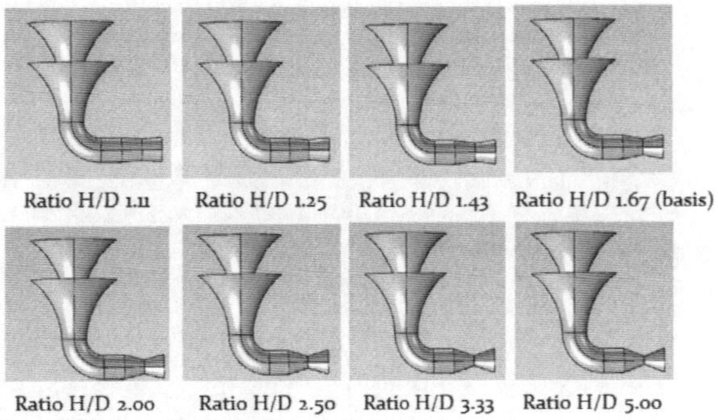

Fig. 15. Ratio of funnel gap variation

3 Result and Discussion

3.1 Result of funnel gap variations

Distribution of wind speed which discussed are variations in ratio Y/X on Free Wind Stream (FWS) 4 m/s. Variations of FWS 1 m/s - 3m/s presented on Table 3.

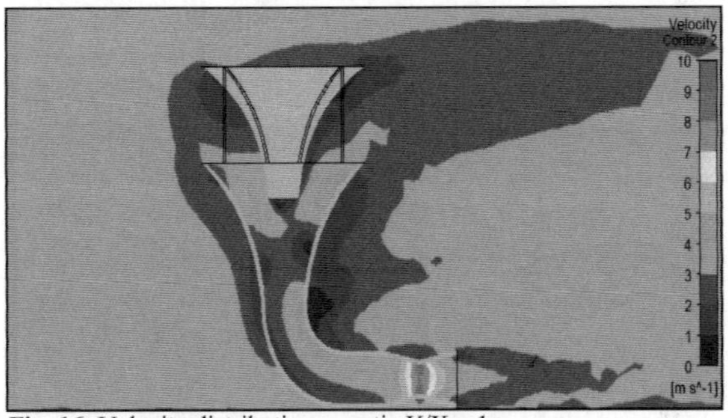

Fig. 16. Velocity distribution on ratio Y/X = 1

From Figure 16, there is variation of the color. Blue color means low speeds and the red color means high

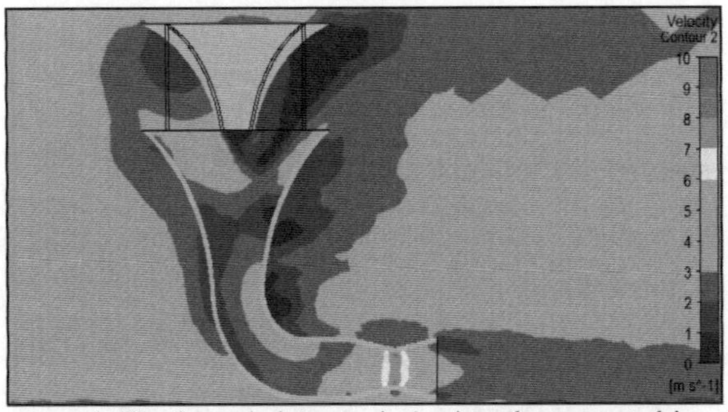

speed. The free wind stream is 4 m/s and represented in blue, while the wind speed at the constriction reaches 8.18 m/s and is represented in orange. The ratio Y/X = 1 is the ratio between the Y and X values in Section II.C.1 where the value Y is 30 feet, and X is 30 feet.

Fig.17. Velocity distribution on ratio Y/X = 0.75

From Figure 17, there is variation of the color. Blue color means low speeds and the red color means high speed. The free wind stream is 4 m/s and represented in blue, while the wind speed at the constriction reaches 9.03 m/s and is represented in red. The ratio Y/X = 0.75 is the ratio between the Y and X values in Section II.C.1 where the value Y is 22.5 feet, and X is 30 feet.

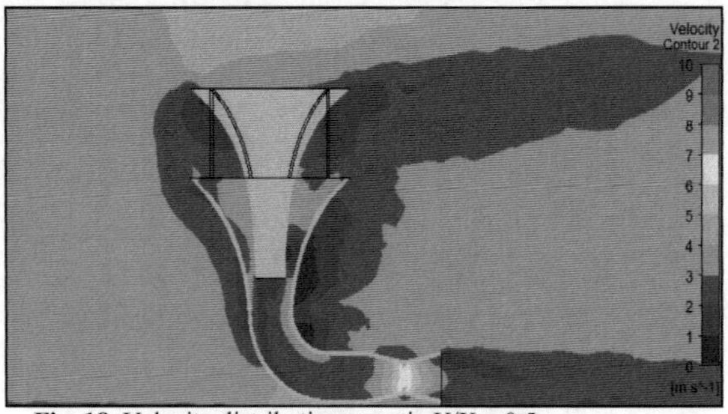

Fig. 18. Velocity distribution on ratio Y/X = 0.5

From Figure 18, there is variation of the color. Blue color means low speeds and the red color means high speed. The free wind stream is 4 m/s and represented in blue, while the wind speed at the constriction reaches 7.45 m/s and is represented in light orange. The ratio Y/X = 0.5 is the ratio between the Y and X values in Section II.C.1 where the value Y is 15 feet, and X is 30 feet.

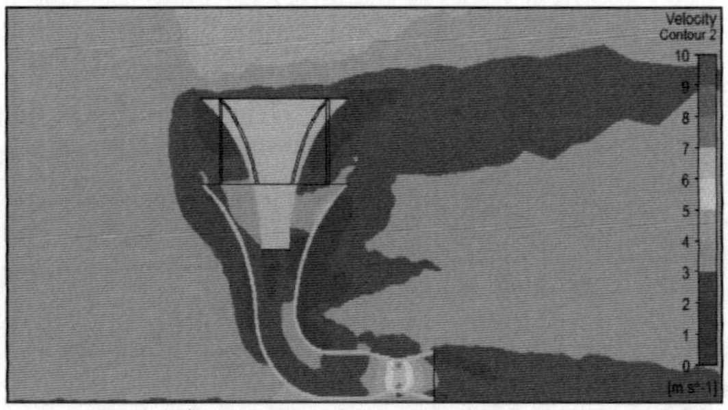

Fig. 19. Velocity distribution on ratio Y/X = 0.25

From Figure 19, there is variation of the color. Blue color means low speeds and the red color means high speed. The free wind stream is 4 m/s and represented in blue, while the wind speed at the constriction reaches 7.3 m/s and is represented in yellow. The ratio Y/X = 0.25 is the ratio between the Y and X values in Section II.C.1 where the value Y is 7.5 feet, and X is 30 feet.

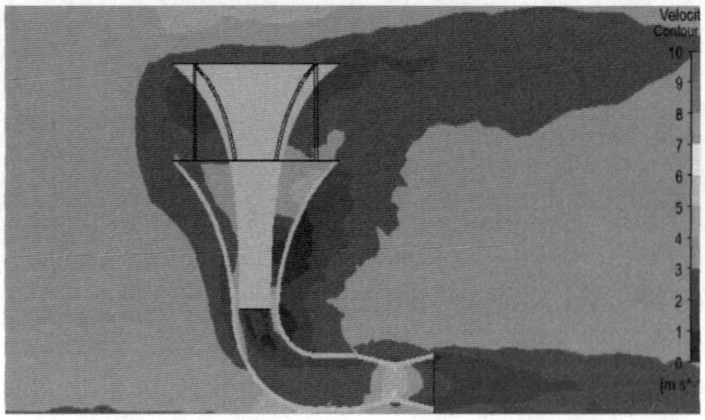

Fig.20. Velocity distribution on ratio Y/X = 0

From Figure 20, there is variation of the color. Blue color means low speeds and the red color means high speed. The free wind stream is 4 m/s and represented in blue, while the wind speed at the constriction reaches 5.16 m/s and is represented in green. The ratio Y/X = 0 is the ratio between the Y and X values in Section II.C.1 where the value Y is 0 feet, and X is 30 feet.

This research analyze the average wind speed at the throat. The following describes the average wind speed at each ratio Y / X and Free Wind Stream.

Table 3. Simulation result of each ratio Y/X variation and FWS

Simulation Results					
Ratio Y/X	0	0.25	0.5	0.75	1
FWS = 1 m/s					
Wind speed at throat (m/s)	1.29	1.83	1.86	2.26	2.05
Speed Ratio	1.29	1.83	1.86	2.26	2.05
FWS = 2 m/s					
Wind speed at throat (m/s)	2.58	3.65	3.72	4.51	4.09
Speed Ratio	1.29	1.83	1.86	2.26	2.05
FWS = 3 m/s					
Wind speed at throat (m/s)	3.87	5.47	5.58	6.77	6.14
Speed Ratio	1.29	1.83	1.86	2.26	2.05
FWS = 4 m/s					
Wind speed at throat (m/s)	5.16	7.3	7.45	9.03	8.18
Speed Ratio	1.29	1.83	1.86	2.26	2.05

Table 3 is the data results of the variations in the ratio Y/X as well as free wind stream variation to the speed at the throat obtained using ANSYS 16.2 is shown in a graph such as Figure 21.

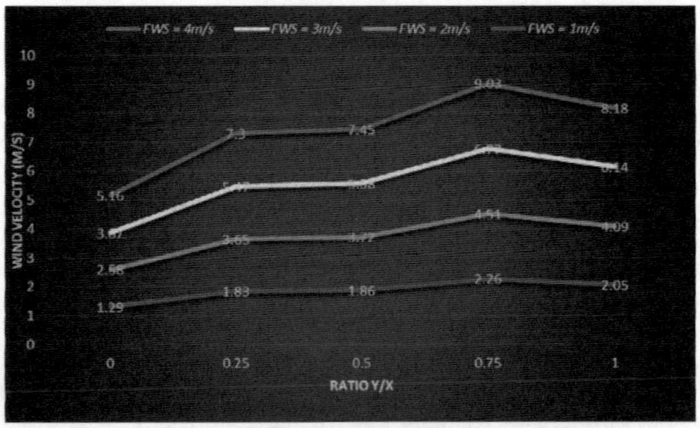

Fig. 21. Graphicof effect of Y/X ratio to wind velocity at venturi

From the research that have been conducted using ANSYS 16.2, it was found that the optimal ratio of funnel gap (Y) and length of the funnel (X) at low wind speed variation is 0.75. Increase the wind speed reached 2.26 times to the free wind stream with each speed detail in Table 3.

3.2 Results of throat diameter variations

Distribution of wind speed which discussed are variations in ratio H/D on Free Wind Stream (FWS) 4 m/s. Variations of FWS 1 m/s - 3m/s presented on Table 4.

https://doi.org/10.1051/e3sconf/20184201013

Fig. 22. Velocity distribution on ratio H/D = 1.11

Figure 22 shows the distribution of wind speeds where blue indicates low speed and red colors indicate high speed. The throat diameter is 9 feet. The average wind speed at the throat is 3.74 m/s.

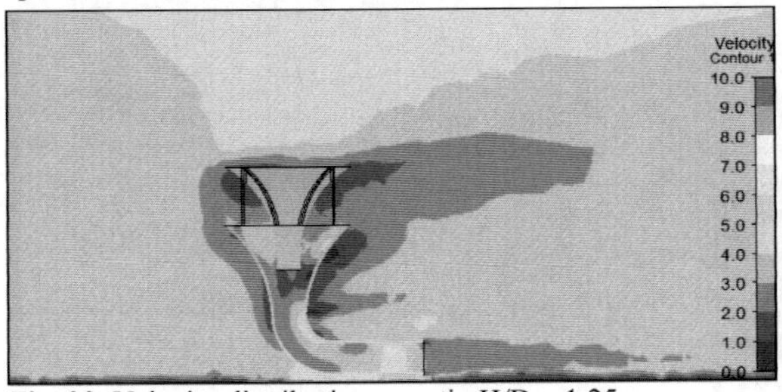

Fig. 23. Velocity distribution on ratio H/D = 1.25

Figure 23 shows the velocity distribution in the throat 8 feet diameter. An increase in the speed of previous variations are marked in light green color. The average wind speed at the throat is 4.80 m/s.

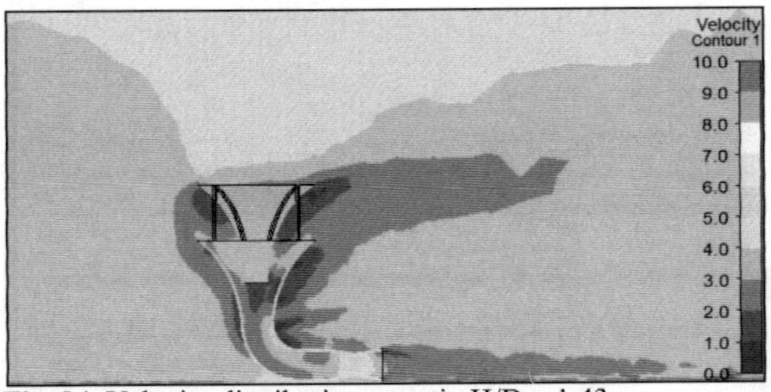

Fig. 24. Velocity distribution on ratio H/D = 1.43

Figure 24 shows the velocity distribution in the throat 7 feet diameter. An increase in the speed of previous variations are marked in green color. The average wind speed at the throat is 6 m/s.

Fig.25. Velocity distribution on ratio H/D = 1.67

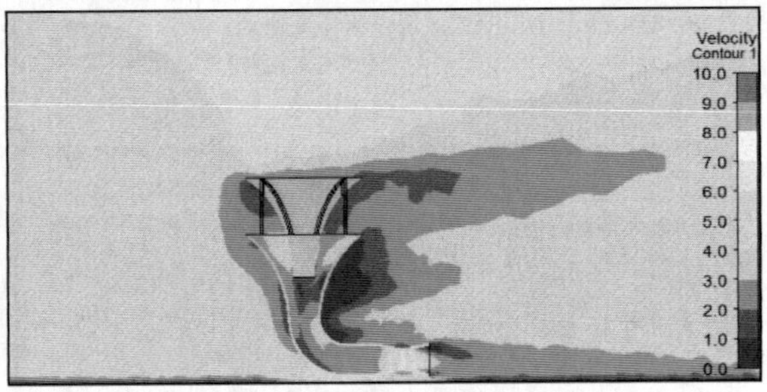

Figure 25 shows the velocity distribution in the throat 6 feet diameter. This ratio is the ratio of the base used for validation. An increase in the speed of the previous variation that is characterized by the color of the throat is more yellow. The average wind speed at the throat is 7.45 m / s.

Fig. 26. Velocity distribution on ratio H/D = 2.00

Figure 26 shows the velocity distribution in the throat 5 feet diameter. An increase in the speed of previous variations are marked in red color. The average wind speed at the throat is 9.23 m/s.

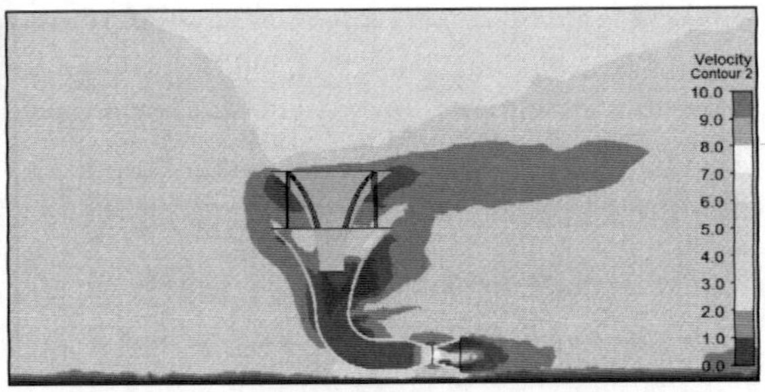

Fig. 27. Velocity distribution on ratio H/D = 2.50

Figure 27 shows the velocity distribution in the throat 4 feet diameter. There is a speed reduction from the previous variation that is characterized by the color of the throat back in yellow after the red in the previous variations. This shows that the decrease in the value of

 https://doi.org/10.1051/e3sconf/20184201013

the diameter of the esophagus is not always better at increase wind speed. The average wind speed at the throat is 8.22 m / s.

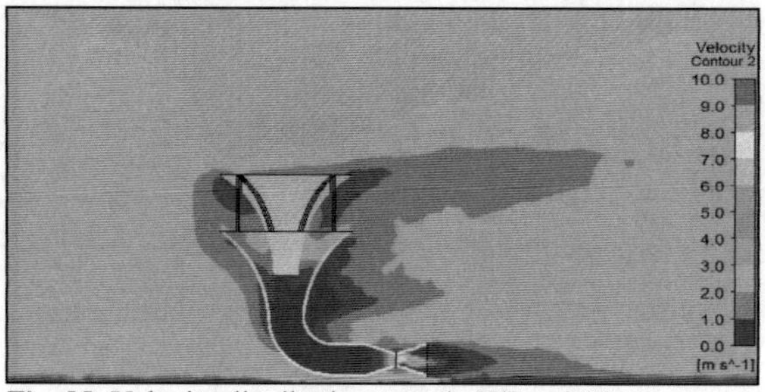

Fig. 28. Velocity distribution on ratio H/D = 3.33

Figure 28 shows the velocity distribution in the throat 3 feet diameter. A decrease in the speed of previous variations are marked in yellow color. The average wind speed at the throat is 7.44 m/s.

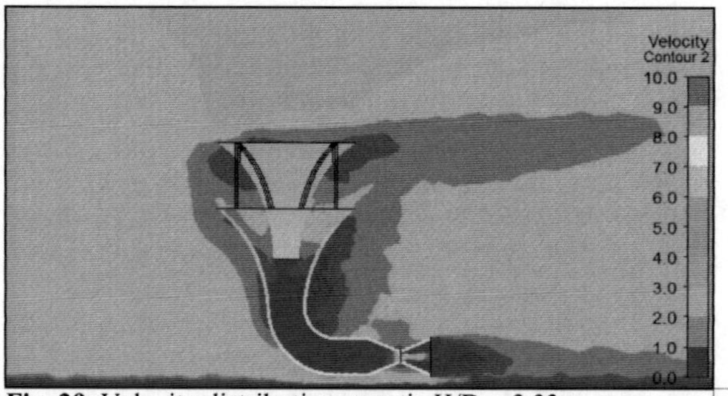

Fig. 29. Velocity distribution on ratio H/D = 3.33

Figure 28 shows the velocity distribution in the throat 2 feet diameter. A decrease in the speed of previous variations are marked in blue color. The average wind speed at the throat is 2.09 m/s.

Table 4. Simulation result of each ratio H/Dand FWS

Simulations using ANSYS								
Ratio H/D	1.11	1.25	1.43	1.67	2.00	2.50	3.33	5
Free Stream Wind 4 m/s								
wind speed at throat (m/s)	3.74	4.80	6.00	7.45	9.23	8.22	7.44	2
wind speed ratio	0.94	1.20	1.50	1.86	2.31	2.06	1.86	0
Free Stream Wind 3 m/s								
wind speed at throat (m/s)	2.82	3.48	4.50	5.58	6.94	5.76	5.58	1
wind speed ratio	0.94	1.16	1.50	1.86	2.31	1.92	1.86	0
Free Stream Wind 2 m/s								
wind speed at throat (m/s)	1.86	2.40	3.01	3.72	4.60	3.78	3.71	1
wind speed ratio	0.93	1.20	1.51	1.86	2.30	1.89	1.86	0
Free Stream Wind 1 m/s								
wind speed at throat (m/s)	0.94	1.15	1.40	1.85	2.30	1.89	1.85	0
wind speed ratio	0.94	1.15	1.40	1.85	2.30	1.89	1.85	0

Table 4 is the data results of the variations in the ratio H/D as well as free wind stream variation to the speed at the throat obtained using ANSYS 16.2 is shown in a graph such as Figure 30.

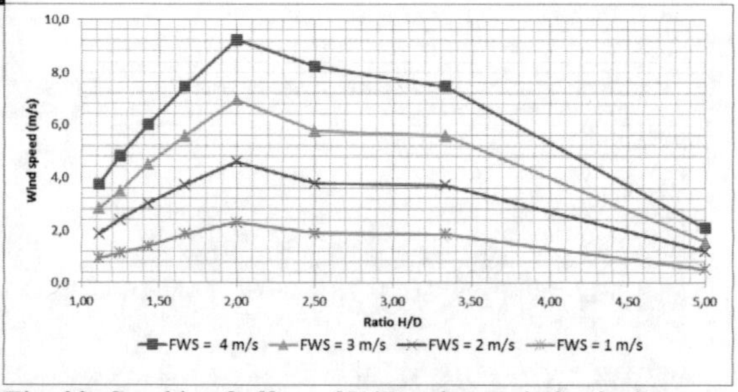

Fig. 30. Graphic of effect of Y/X ratio to wind velocity at venturi

From the research that have been conducted using ANSYS 16.2, it was found that optimal ratio of the diameter of inlet venturi (H) and throat diameter (D) at low wind speed variation is 2. Increase the wind speed reached 2.31 times to the free wind stream with each speed detail in Table 4.

3.3. Comparing with the Existing Design

The result on Secntion III.A and III.B will be compared to the existing design on the reference paper [1] and the data obtained as shown in Table 5.

Table 5. Comparison of basic with optimal design

Design	Wind velocity at venturi	Speed Ratio
Basic	7.45 m/s	1.86
Ratio Y/X optimal	9.03 m/s	2.26
Ratio H/D optimal	9.23 m/s	2.31

The existing design INVELOX has increased the wind speed by 1.86 times, while this research increased the wind speed in the optimal ratio Y/X and H/D by 2.26 times and 2.31 times. Comparing to the reference that the increased of the wind speed in the optimal ratio Y/X and H/D better 21.5% and 24.2% than the existing design.

3 Conclusion

The highest wind speeds occur at a ratio Y/X 0.75 for variation of funnel gap and the ratio H/D 2.00 for the variation of throat diameter. The wind speed reached 2.26 times faster for variation of funnel gap and 2.31 faster for variation of throat diameter. This design is 21.5% better than the basic design for the variation of funnel gap and 24.2% better than the basic design for the variation of throat diameter.

References

1. D. Allaei, Y. Andreopoulos, *Energy*, **69**, 336-344 (2014)
2. D. Allaei, Y. Andreopoulos, "INVELOX: a new concept in wind energy harvesting". In: ASME 2013 7th international conference on energy sustainability & 11th fuel cell science, engineering and technology conference ES-fuel cell, Minneapolis, MN, USA. July 14-19, (2013)
3. D. Allaei, Using CFD to predict the performance of innovative wind power generators, *Proc. of the 2012 COMSOL Conf.*, Boston, (2012)
4. L. Chen, F.L. Ponta, L.I. Lago, *Energy. Sustain. Dev.*, **15**, 398-410 (2014)
5. N.M. Kumar, M.S.P. Subathra, O.D. Cota, *Proc. Tech* **21**, 33-40 (2015)
6. M. Johanson, Evaluation of RANS turbulence models for the hydrodynamic analysis of an axisymmetric streamlined body with special consideration of the velocity distribution in the stern region, *Master Thesis*, KTH, School of Eng. Sciences (SCI), Aeronautical and Vehicle Eng., (2012)
7. I.B. Celik, Introductory Turbulence Modeling, *Lecture Notes*, Mechanical & Aerospace Engineering Dept., West Virginia University, 26506-6106 (1999)
8. C&F Green Energy, CF 20 C&F Green Energy Brochure, http://www.cfgreenenergy.com/cf-20/, April 21, (2016)
9. ANSYS, Inc., ANSYS Workbench User's Guide. November, (2009)
10. ANSYS, Inc., ANSYS Fluent Theory Guide, November, (2013)
11. ANSYS Inc., Introduction to ANSYS Meshing, March (2016)
12. GEF, WhyPGen, BPPT, UNDP, Potential Wind Power Generations in Indonesia, Accessed from http://whypgen-bppt.com/id/, April 21, (2016)
13. ANSYS Inc., Introductory FLUENT Training, December, (2006)
14. A. Bakker, Lecture 7 – Meshing: Applied Computational Fluid Dynamics. Accessed from http://www.bakker.org, June 1, (2016)
15. H.K. Versteeg, W. Malalasekera, *An Introduction to Computational Fluid Dynamics the Finite Volume Method*, Longman Scientific & Technical, Inc., New York, (199

Nuclear Security Training: A New Approach for Educating Nuclear Technology in Indonesia

Ferdiansjah[1], *Susetyo* Hario Putro[1], *Sihana*[1], *Widya* Rosita[1], *Faridah*[1]

[1]Department of Nuclear Engineering and Engineering Physics, Universitas Gadjah Mada, Yogyakarta-55281, Indonesia

Abstract— Indonesia has planned to erect Nuclear Power Plant for along time, but the government always postponed the plan due to the presence of anti-nuclear group that decreases the public acceptance. In addition, the public knowledge on nuclear technology is still low, since they only get in the last month before finish their study in high school. The socialization of nuclear technology done by government body could not significantly increase the public acceptance, since the knowledge and confidence on nuclear technology among the government employee is also low. The recent issue regarding to nuclear security give a new opportunity in socializing strategy. Department of Nuclear Engineering and Engineering Physics Universitas Gadjah Mada (UGM) as the only one university in Indonesia offering nuclear engineering education had conducted several nuclear security training for specific people who had been not targeted in previous socialization programs. The training done in collaboration with National Nuclear Energy Regulatory Body (BAPETEN) of Indonesia was conducted in Batam, Surabaya and Jakarta. Almost 20 people coming from police department, national coast guard, custom agency, etc attended in the training. Training materials consist of introduction on nuclear technology, introduction on nuclear security, physical protection system, radiation detection, transport security, etc. Pre-test and post-test were done to know their improvement. This was first time for most of participants to get nuclear technology knowledge. Based on the survey their knowledge was significantly improved. They felt that the training was valuable for them to know the benefit and risk of nuclear technology. We believe that it could increase the public participant and acceptance on utilizing nuclear technology in Indonesia.

Keywords— *education, nuclear security, training*

NOMENCLATURE

UGM - Universitas Gadjah Mada
BAPETEN - National Nuclear Energy Regulatory Body of Indonesia
BATAN - National Nuclear Energy Agency of Indonesia
I-CoNSEP - Indonesian Centre of Excellence of Nuclear Security and Emergency Preparedness
CSCA - Center for Security Culture and Assessment
CITS - Center for International Trade and Security
UGA - University of Georgia, USA
NSS - Nuclear Security Series
PDC - Professional Development Course
KCL - King's College London
RPM - Radiation Portal Monitor
PRD - Personal Radiation Detector
RID - Radionuclide Identification Detector
NSD - Neutron Search Detector
ITDB - International Threat Data Base

1 Introduction

Indonesia has planned to build Nuclear Power Plant for along time, but the government always postpones the plan due to the presence of anti-nuclear group that decreases the public acceptance. In addition, the public knowledge on nuclear technology is still low, since they only get in the last month before finish their study in high school. The socialization of nuclear technology done by government body could not significantly increase the public acceptance, since the knowledge and confidence on nuclear technology among the government employee is also low. Susetyo Hario Putero, Haryono Budi Santosa and Widya Rosita (2012) said that proactively improving a community's knowledge about nuclear technology is a one of key success factor to opening a public dialogue on the first NPP in Indonesia [1]. They found that nuclear energy technology is only introduced, as part of the radiochemistry topic, in last two months before a student graduates from senior high school. In addition, most of the engineering and science education programs in higher education level are unrelated to nuclear science and technology (more than 96%).

The recent issue regarding to nuclear security addressed by IAEA under IAEA Nuclear Security Plan 2010—2013 give a new opportunity in socializing strategy for specific people who had been not targeted in previous socialization programs. This paper would like to describe the new approach of Department of Nuclear Engineering and Engineering Physics Universitas Gadjah Mada (UGM) as the only one university in Indonesia offering nuclear engineering education to socialize nuclear technology in Indonesia.

2 Current nuclear technology in indonesia and its threats

Since 1965, nuclear technology is known and utilized nuclear technology in Indonesia, especially for medical, food and industrial purposes. But, until know, the government is always postponed their plan to build the first NPP due to the protester, eventhough in national energy policy for 2006-2025, NPP is planned to supply about 2% of the national energy demand. It is believed that most of Indonesian has no adequate knowledge of nuclear technology got from formal education as mentioned before. Their knowledge on "nuclear" word is mostly influenced by several negative and improper things, such as the atomic bomb, human radiation experiments, and the genetic mutations pictured as "monster" or "superhero" in popular comics.

For increasing the knowledge, institutions related to nuclear technology (BAPETEN, National Nuclear Energy Agency of Indonesia/BATAN, UGM, etc,) have conducted several activities, such as educating community leaders, seminars, "Atom day", community emporwent, etc. UGM itself has four targets in educating people, i.e. schools student, rural people, urban communities and scientific community. There were some positive impacts from the target groups [2]. Their knowledge on risks and benefits of nuclear technology is more balance.

The recent issue in utilization of nuclear technology is nuclear security. Because it is still new issue in Indonesia and hence the knowledge for securing radioactive material is questioned. In the other hand, several terrorist acts had attacked people in Indonesia. Last January, suicide bombing and gun firing done were attacked Jakarta and killed seven people including the terrorists. Eventhough, radionuclide did not present in the attacks, but the possibility could not be neglected.

Nuclear security issue is not only related to the terrorism, but also could be associated to the international nuclear materials transport either sea or air transports. Since Indonesia is a strategic position between Asia and Australia, Pacific Ocean and Hindia Ocean, the risk in nuclear traffic and trade from one country to other through Indonesia territorial is high. Security risk could also be escalated resulted from the escalation of regional security risk. OECD/IEA said that there are 6 power reactors proposed to be erected in Southeast Asia countries region to supply their electricity demands [3].

For overcoming the security issue, BAPETEN has initiated Indonesian Centre of Excellence of Nuclear Security and Emergency Preparedness (I-CoNSEP) for coordinating all activities related to the nuclear security. UGM has signed a Memorandum of Undertanding with BAPETEN to serve as a national centre for nuclear security education coordinated under I-CoNSEP [4]. In order to build national nuclear security capacity, UGM actively promotes and educates nuclear security to community by maximizing internal resources and external networking.

BATAN actively builds the security culture among their staffs as well as safety culture. They also conducted nuclear security culture self-assessment based on IAEA methodology as the first pilot project in the world in 2012. It was supported by IAEA and experts from CITS-UGA, USA. BATAN has decided to establish a Center for Security Culture and Assessment (CSCA) on 29 September 2014, as an expert and training hub nationally, regionally and internationally in cooperation with the IAEA and CITS-UGA [5]. The center succesfully conducted the second self-assessment in 2015.

3 Nuclear security training in indonesia and its impacts

The committment of UGM to overcome nuclear security issue is an implementation of the vission to become one of center for excellent in nuclear engineering education in South East Asia region. To realize it, UGM has conducted international nuclear security education professional development course (PDC) based on NSS-12 in collaboration with Centre for Science & Security Studies, King's College London (KCL) in 2014 and 2015. Based on the experience in running PDC, UGM conducted some nuclear security trainings in collaboration with BAPETEN to build national nuclear security capacity.

There were two type of trainings. The first was four days training to train Front Liner Officer (Harbor Officer, Police Officer and National Ocean Guard Officer) in two big harbors (Batam and Surabaya). Some hospital staffs, custom officers and East Java Disaster Management Authority staffs also joined in Surabaya training. The second was one day training for management team who will make some rules or decisions in nuclear area.

The materials for first type were not only nuclear security related materials (introduction on nuclear security, physical protection system, transport security, etc.), but also consisted of introduction on nuclear technology (basic of ionizing radiation, radiation effect, radiation units, safety principle and radiation detection) in order to build a strong foundation in understanding nuclear security. Materials were delivered using multiple delivery methods that are classical, field visit, case study, watching related videos, practical works and table top exercise to keep the enthusiasm of trainees.

The participants in second type training were higher profile persons than the first type. Therefore, materials were related to the security policy, international and national regulations, and introduction to I-CoNSEP that were delivered by using classical method. Pre-test and post-test were done to know their improvement by comparing the knowledge before and after attending training.

4 Research Methodology

The impact of the training in nuclear security will be measured by two procedures :

1. Pre test question
2. Post test question

Pretest questions are intended to gain information about existing knowledge of the participants for several topics related to nuclear security. This information is very important so in the future the appropriate level of training could be tailored.

Post test questions could be used to measure the impact of training for the participants. The data is very important because it could measure the increase of knowledge after training is completed.

The score that is used to measure the level of knowledge varies from 0 to 7, where 0 denotes no prior knowledge of the subject and 7 shows the full knowledge regarding with the subject. Impact could be determined by comparing an increase on the level of knowledge to the initial level of knowledge

5 Results and discussion

Survey in first type training shows that about 20% of trainees had already haven basic knowledge on nuclear security. They already known how to use radiation detector. But, only 6% of 26 trainees in Batam and 15% of 17 trainess in Surabaya had a good understand. Therefore, the training was the first event to know not only nuclear security but also nuclear technology for the most participants, although dealing with radiation should be a part of their job by national regulations. They looked enthusiastic to learn and want to attend in next training. Survey in second type training for higher profile people confirmed that advanced national nuclear security training is needed.

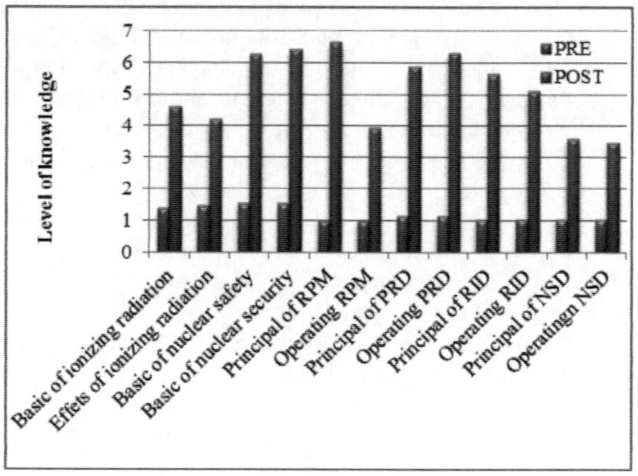

(a)

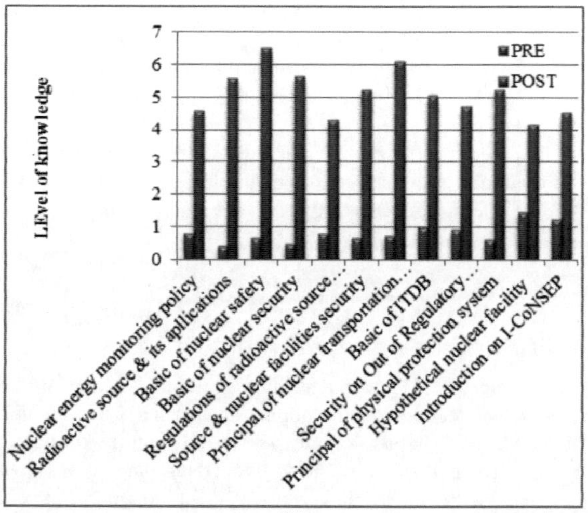

(b)

Figure 1. Pre-test and post-test results in (a) Batam and (b) Surabaya.

Fig. 1 shows that their knowledge was significantly improved from below level 2 to upper level 4 (average). In some materials (basic safety, basic security, principal of radiation detector, principal of physical protection, and principal of security in transport) the improvement reached more than level 6. It was believed that the delivery method that were used for those topics (case study, practical work and table top exercise) affected the improvement. The results in Surabaya was better than Batam one, because there was some revision on delivery method based on Batam's experience. The improvements on the knowledge of Neutron Search Detector (NSD) were lowest, because there was no neutron source for practicing, and hence the trainees was difficult to understand. But, in general, we could say that trainings succesfully improved their knowledge on nuclear security.

Discussion during training, either in class or break, was not only limited to nuclear security, but also nuclear technology in common, especially that related to nuclear safety in their job activities. Nuclear safety and waste management were favorite topics to be discussed. The important thing in training is how the trainers could improve their self-confidence in dealing with radioactive materials by knowing the benefit and risk of nuclear technology. We believe that it could increase the public participant and acceptance on utilizing nuclear technology. Comments and impressions from participants written in post-test show that the training was valuable for them. It is proven when trained National Ocean Guard Officers in Batam in collaboration with BAPETEN arrested an international ship which presumed to bring illegal uranium across Batam area.

6 Conclusions

Nuclear security training was first time for most of participants to get nuclear technology knowledge. Based on the survey their knowledge was significantly improved. They felt that the training was valuable for them to know the benefit and risk of nuclear technology. We believe that it could increase the public participant and acceptance on utilizing nuclear technology in Indonesia.

Acknowledgment

The authors gratefully acknowledge the support of the Department of Nuclear Engineering and Engineering Physics, Universitas Gadjah Mada. We also thank BAPETEN for cooperation. Support from Partnership for Nuclear Security (PNS)-USA and CRDF Global in presenting this paper are acknowledge.

References

[1] Susetyo Hario Putero, Haryono Budi Santosa and Widya Rosita, 2012, Challenge in the Public Acceptance and Cooperation on the Verge of Building the First Nuclear Power Plant in Indonesia, *Proc. of the 2012 20th International Conference on Nuclear Engineering Collocated with the ASME 2012 Power Conference,* ASME, Anaheim.

[2] Widya Rosita, Anung Muharini, Ester Wijayanti, Susetyo Hario Putero, 2015, *The Role of Universitas Gadjah Mada on Forming Public's Perception about Nuclear in Indonesia through Community Empowerment Programs,* Paper presented in 2015 Women in Nuclear Annual Meeting, Vienna.

[3] *Asia's Nuclear Energy Growth*, Available on http://www.world-nuclear.org/ (accessed on 2nd April 2014).

[4] Susetyo Hario Putero, et.al., 2014, The Role of Universitas Gadjah Mada in the Recent Global Nuclear Security, *Proc. of the 3rd Applied Science for Technology Application, ASTECHNOVA 2014 International Energy Conference,* Yogyakarta.

[5] Anhar R. Antariksawan and Khairul, 2015, *Nuclear Security Culture and Assessment: BATAN's Experience*, Paper presented in ASTECHNOVA 2015 International Energy Conference, Yogyakarta.

Development of Fuel Debris Inspection Methods using Air-Coupled Ultrasound

Ari Hamdani[1*], *Hiroshige* Kikura[1], *Shun* Kimura[1], *Keisuke* Tsukada[1], *Daisuku* Sasa[2], and *Shuichi Oomori*[2]

[1]Laboratory for Advanced Nuclear Energy Tokyo Institute of Technology Tokyo, Japan
[2]Tokyo Electric Power Company Holding (TEPCO) Yokohama, Japan

Abstract. Japan has decided to start fuel debris removal on 2021. Therefore, before starting retrieval of the fuel debris, investigation of location, shape, and also properties of fuel debris have to be performed. In this study, air-coupled ultrasound technique had been developed in order to visualize of fuel debris. Firstly, simulated fuel debris was represented by stone and experiment was done in laboratory scale. Flat and focused air-coupled ultrasound were used as a sensor to get echo signals of the target surface. Results showed that both flat and focused sensor successfully reconstructed the flat surface image in good spatial resolution. However, flat air-coupled transducer shown unstructured. On the other hand, focused air-coupled ultrasound shown good performance when target surface was unstructured. Therefore, it was recommended to used focused sensor when target surface was unstructured like simulated fuel debris.

1. Introduction

On March 11, 2011, a massive earthquake and tsunami struck Tohoku area of Japan. As a consequence, there was a severe accident on Fukushima Daiichi Nuclear Power Plant (NPP). This accident released a significant quantity of radionuclides into the environment and three of the plant's six reactors suffered core meltdowns (unit 1,2 and 3). A report by Tokyo Electric Power Company (TEPCO) [1], in unit 1 almost all of melted fuel has fallen down to the Primary Containment Vessel (PCV) and little fuel has left in Reactor Pressure Vessel (RPV). For unit 2 and 3, some part of melted fuels has fallen down to the bottom of RPV plenum and PCV pedestal and the other part may have fallen left inside RPV. Recently, fuel removal completed in unit 4 on December 22, 2014. Hence, retrieval of the debris, especially in unit 1, 2 and 3, is the primary requirement for the ending, of the accident. However, the way toward retrieval of debris is long tough, In mid-and-long-term decommissioning roadmap by TEPCO, fuel debris retrieval is expected to begin on 2021 [2].

Therefore, efficient and reliable technology for removal fuel debris is indispensable. First of all, the location, the shape, and also the properties of debris have to be investigated, while it is continuing refusing human's approach by generating extremely high-level radiation. Currently, researchers have identified two methods for retrieving highly radioactive nuclear fuel debris in three reactors at Tokyo Electric Power Company's (Tepco's) Fukushima Daiichi plant [3]. They are submersion method and dry method as shown in Figure 1. Before perfoming a retrieval fuel debris, the shape of fuel debris in RPV and PCV must be visualized. A nondestructive inspection method for identifying the position and shape of fuel debris is indispensable. The novel measurement technique for fuel debris visualization mush have these features [3]:

1. Enable remote control to keep low exposures for workers.
2. Avoid interference with existing structures.
3. Adapt to operations in air (high dose and high humidity).
4. Enable long continuous use.

A previous study, Takegami et al. [4] investigated internal visualization of the High-Temperature Engineering Test Reactor (HTTR) of Japan Atomic Energy Agency by using cosmic-ray muons. Such as concrete walls of the containment vessel of the HTTR. From the results of the preliminary examination, it appears that the inspection method with muons is promising for searching for fuel debris in a reactor. However, another measurement technique must be developed when muon technique cannot be used in good order. As shown in Figure 1.b., the novel measurement technique must adapt to operations in the air with high dose radiation and humidity.

When the dry method is used a retrieving fuel debris, ultrasound technique can be a good candidate for visualization mehtod. Generally, ultrasound technique needs a coupling medium such as water for better ultrasound's signal propagation. Nevertheless, ultrasound imaging in the air is advatageous because the position of the transducers can be set freely. Furthermore, imaging in the air provides higher resolution than in water because the sound wavelength is shorter in air at a given frequency [5]. Fox *et al.* [6] successfully demonstrated ultrasonic images visualization by 2 MHz using an air-coupled transducer. Consequently, based on previous studies, a measurement technique for fuel debris inspection is developed in our laboratory. The objective of this study is to visualize the shape of fuel debris surface in laboratory scale. A preliminary experiment is conducted at room temperature and a stone represents as simulated fuel debris.

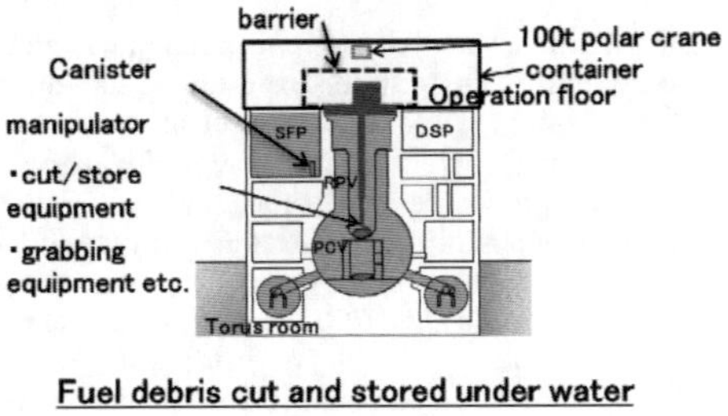

a. Submersion method

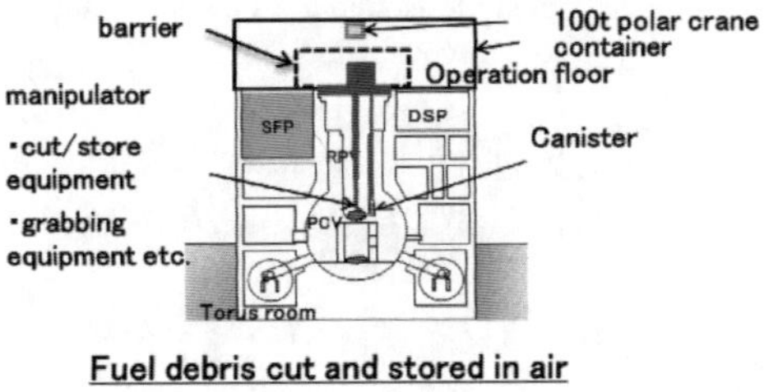

b. Dry method

Fig. 1. Fuel debris retrieval methods [3].

2. Experimental Apparatus

2.1 Air-coupled ultrasound transducer

Generally, ultrasound technique needs a coupling medium such as water for better ultrasound's signal propagation. Air-coupled ultrasound is a non-contact method, thus the difference of acoustic impedance must be taken care. The transmission of two media with high difference acoustic impedance e.g. piezo-electric ceramic (Z_{pc}) and air (Z_a) can be reduced by using a matching layer. Therefore, the ultrasonic wave passes with three different acoustic impedances as shown in Fig.2.

The total energy transmission coefficient is given by

$$T_{t_{total}} = \frac{4.Z_{pc}.Z_a}{\left(\frac{Z_a}{Z_{pc}}+1\right)^2 - \left(\left(\frac{Z_a}{Z_{ml}}\right)^2-1\right).\left(\left(\frac{Z_{ml}}{Z_{pc}}\right)^2-1\right).\sin^2 kd} \quad (1)$$

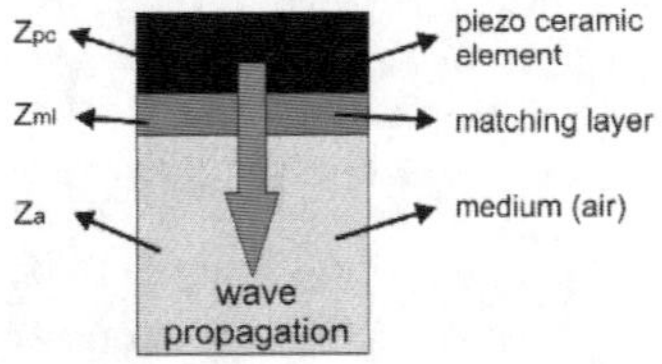

Fig. 2. Impedance matching layer between piezo-electric ceramic and air.

Where Z_{pc}, Z_a and Z_{ml} are the acoustic impedance of piezoelectric ceramic, air and matching layer. K is a wave number of the matching layer and d is matching layer thickness. In this study, flat (0.4K×20N-RX) and focused (0.4K20N) air-coupled ultrasound transducers are used as shown in Fig.3. These two transducers are manufactured by Japan Probe and the basic frequency is 400kHz.

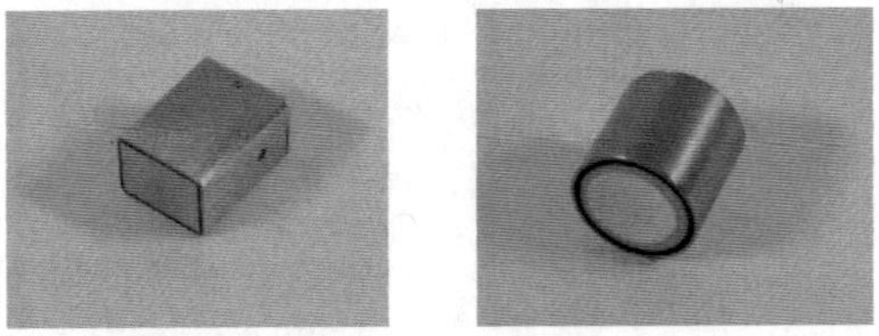

a. flat (0.4K×20N-RX) - sensor A b. focused (0.4K20N) - sensor B

Fig. 3. air-coupled ultrasound transducers.

2.2 Experimental procedure

The basic theory of air-coupled ultrasonic visualization method is simple because it depends mainly on the sound velocity through the air. The air-coupled transducers scans across the surface and sends ultrasound waves to the sample surface, the ultrasound waves are reflected back from the surface along the same path as the incident ultrasound wave. Time-of-flight images of the sample surface are acquired and converted to depth / surface profile images using the simple relation

$$d = \frac{c.t}{2} \quad (2)$$

Where d is distance, t is time-of-flight and c are the velocities of sound in air. In this experiment, three-dimension stage automation is used to move air-coupled ultrasound transducer and visualize the three-dimensional shape of the target. The experiment apparatus is constructed and the simplification experimental apparatus is shown in Figh.4.

Transmission and reception of ultrasonic wave are perfomed by using Japan Probe Co. Pulse-receiver (JPR-10CN). Three-axis stage automation is made by Sigma Koki (OSM26-300 (X) and OSM26-200 (Z)) and controlled by stage controller (SHOT-304GS). Both pulse receiver and three-dimension stage automation are controlled by self-made software as shown in Fig.5.

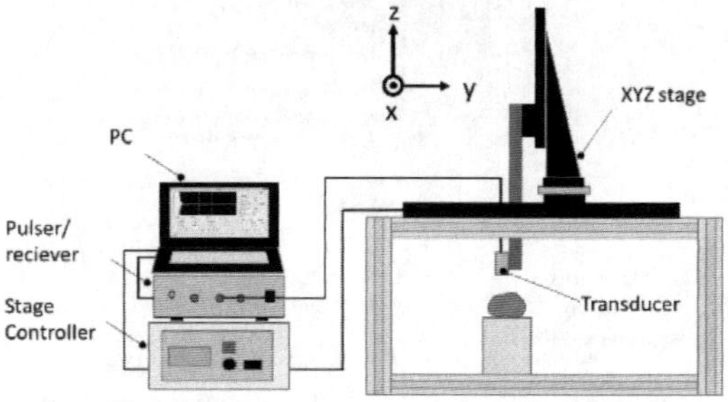

Fig. 4. Air-coupled ultrasound experimental apparatus.

Firstly, pulse-receiver sends an electrical signal to the sensor, then the ultrasonic wave is propagated in the air and reflected back to the sensor by the reflector. The received signal is then converted to a digital signal using AD converter and recorded by the personal computer. The control program has functional as below:

1. Direct control of pulse-receiver e.g. frequency, excitation voltage, wave number, high pass filter, low pass filter, etc.
2. Send and receive an ultrasonic wave and display the received signal.
3. Directly control the stage automation in the x, y, z-direction.
4. Display and storage the echo signals while scanning in the x,y, z-direction.

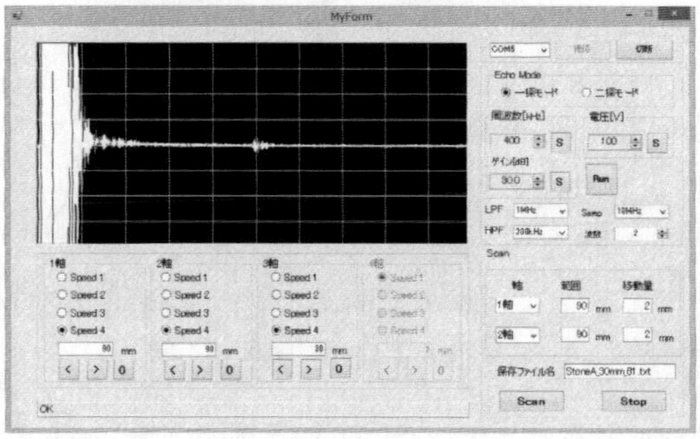

Fig. 5. Device's control program.

Figure 6 and 7 shows the reference reflector and simulated debris are used in this experiment.. A cylindrical stainless steel with 50 mm diameter and 50 mm in height is used as a reference reflector. Simulated fuel debris is represented by a stone with unstructured shape's surface.

Fig. 6. Cylindrical stainless steel's reflector (reference reflector).

Fig. 7. Simulated fuel debri's reflector.

The preliminary test of the echo signal measurement is done for both reflectors. In the experiment, the ultrasound transducer z-axis is moved to the positive direction, to observe the change of the echo signal while changing the distance between the reflector and an ultrasonic transducer. The positional relationship between the sensor and the reflector is shown in Fig.8. The reflector is placed on top of the acrylic block 90 × 90 × 90 mm, the initial distance (*H*) to the center of the reflector is set 10 mm. The air-coupled transducer transmits ultrasonic waves from 10 mm to the target surface, then continued to another position by every 5 mm in the *z* direction. Measurement settings for air-coupled ultrasound are shown in Table 1.

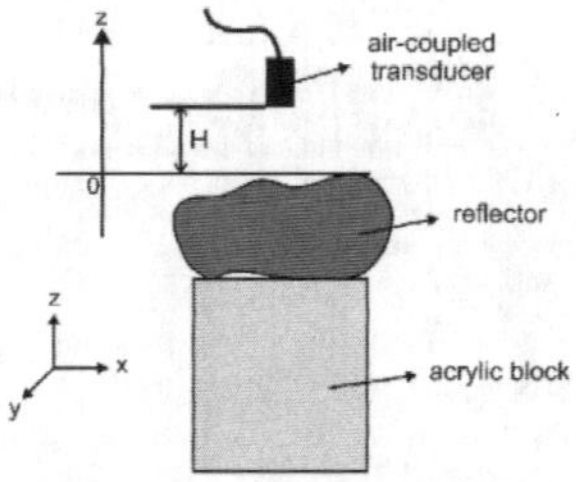

Fig. 8. The position between the sensor and reflector.

Table 1. Measurement Settings

Parameter	Value
Voltage	100 V
Basic frequency	400 kHz
Wave number	2
Gain	30 dB
Low pass filter	1 MHz
High pass filter	200 kHz
Sampling rate	10 MHz

The reconstruction of the two-dimensional image is performed for both reference reflector and simulated fuel debris. Flat transducer sensor (sensor A) and focused sensor (sensor B) are used to acquire echo signals in every 2 mm either x or y-direction. Measurement of the echo signals is automatically controlled by three-dimensional stage automation and the distance of sensor to the reflector's center is determined by echo signals measurement results. Reconstruction image is constructed by seeking time of flight at every measurement position point. Therefore, the distance between sensor and target surface can be obtained by Eq. 1.

3. RESULTS AND DISCUSSIONS

3.1 Echo signals (preliminary experiment)

Figure 9 shows an example of the echo signals coming from the surface of the reference reflector measured by sensor A. It can be seen that echo signals from the cylinder surface can be obtained. Furthermore, it can be confirmed that time of flight and echo signal's intensity change when the distance H is changed in the z direction. The time of flight represents a distance between sensor and target surface. However, at H = 10 and H = 15 mm, the echo signals from target surface are difficult to be distinguished. It is due to near field or dead zone region (below 200 μs) close to transducer's surface. Thus, it is recommended to set initial distance (H) at least 30 mm from the target surface.

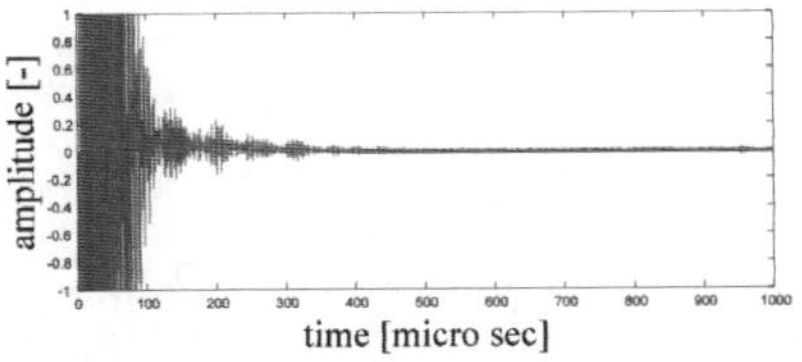

a. H = 10 mm

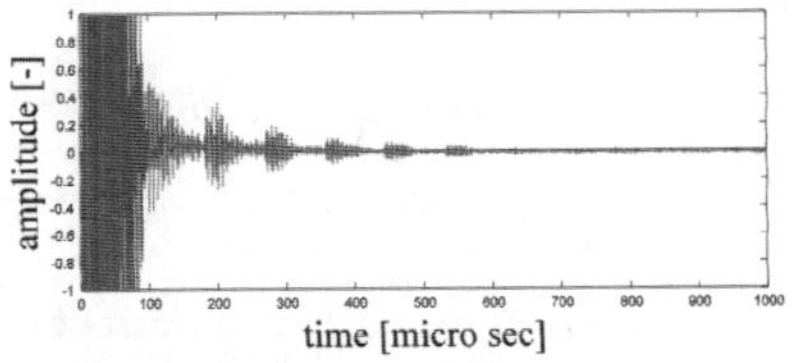

b. H = 15 mm

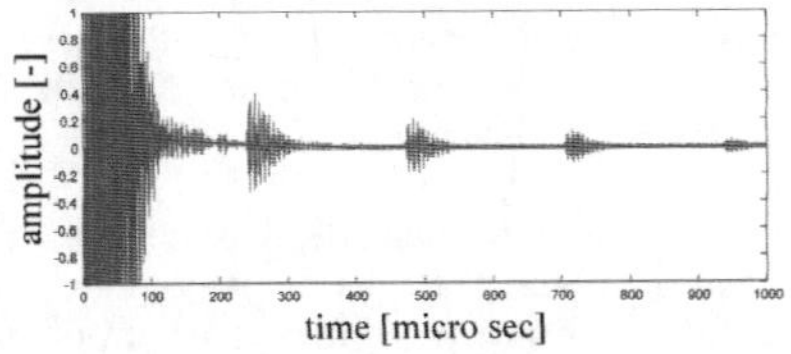

c. H = 40 mm

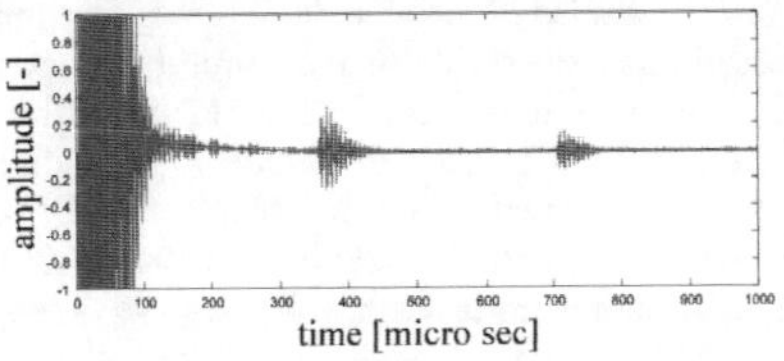

d. H = 60 mm

Fig. 9. Echo signal of the reference reflector measured by sensor B.

Figure 10 shows echo signals coming from reference reflector measured by sensor B. It can be seen that focused air-coupled ultrasound sensor sufficiently obtain echo signals from the target surface. Furthermore, comparing to the results in Figure 9, the echo signal's intensity measured by sensor B is higher. It means that sensor B has better sensitivity compare with sensor A. However, it can be found that dead zone still appears below 100 μs.

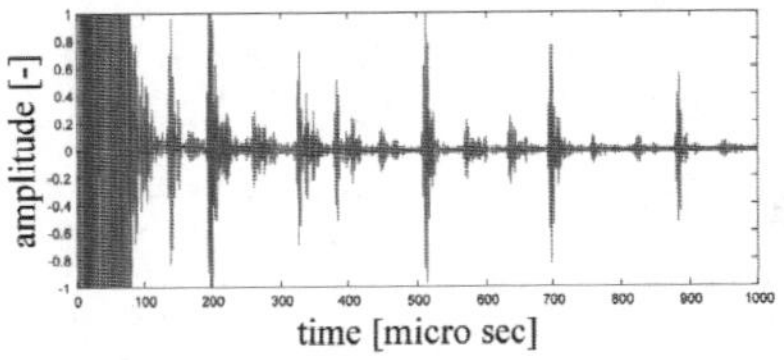

a. H = 10 mm

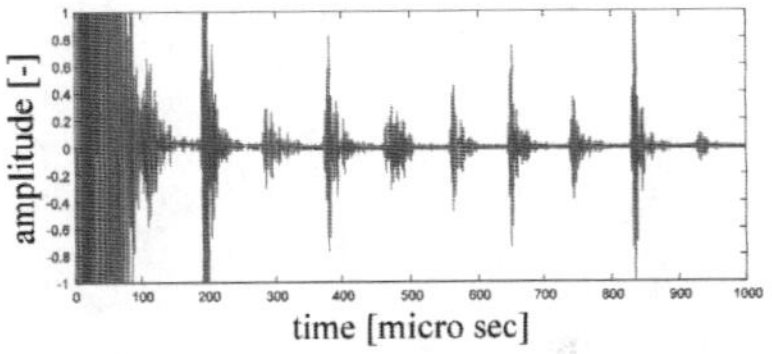

b. H = 15 mm

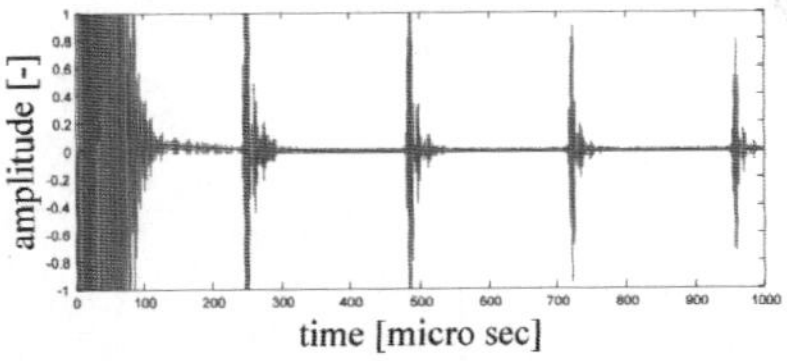

c. H = 40 mm

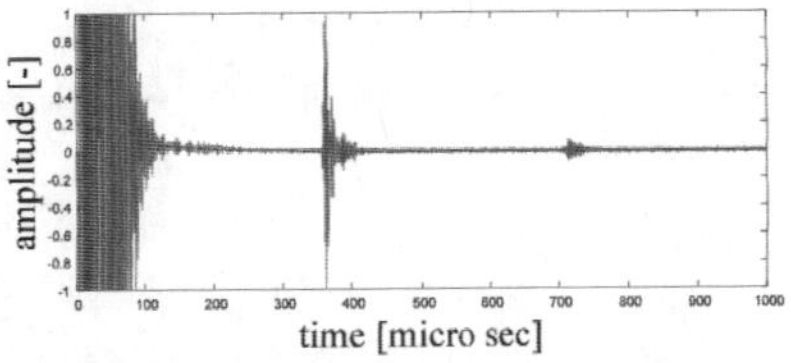

d. H = 60 mm

Fig. 10. Echo signal of the reference reflector measured by sensor B.

Figure 11 and 12 show echo signals coming from simulated fuel debris reflector measured by sensor A and B respectively. It can be seen in Fig. 11 that echo signals cannot be obtained in any distance *H*, it is due to the irregularity of target surface. As a result, neither reflected ultrasound wave is not received by sensor nor attenuated. On the other hand, measured echo signals are possible using sensor B as shown in Fig. 12. Because of sensor B has a curvature on its surface, thus ultrasound wave can be focused into a small area of the target surface. In addition, echo signals which are obtained by sensor B are distinguishable. But still, dead zone (near field region) cannot be avoided.

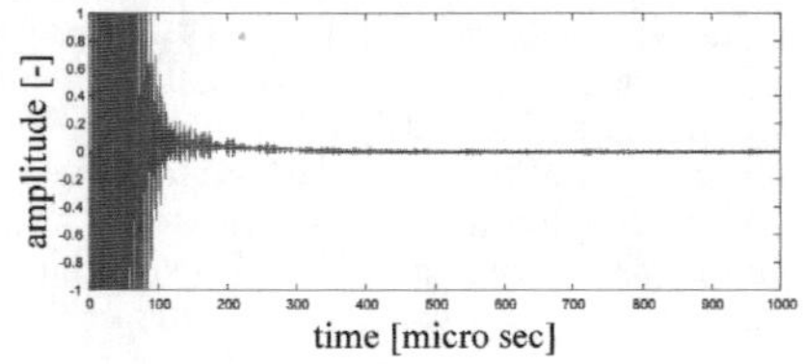

a. $H = 10$ mm

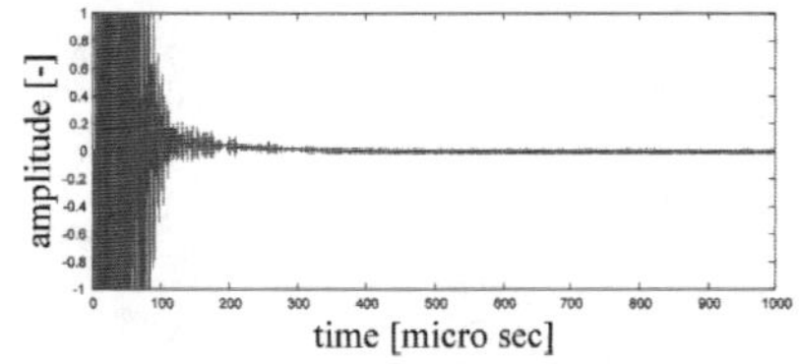

b. $H = 15$ mm

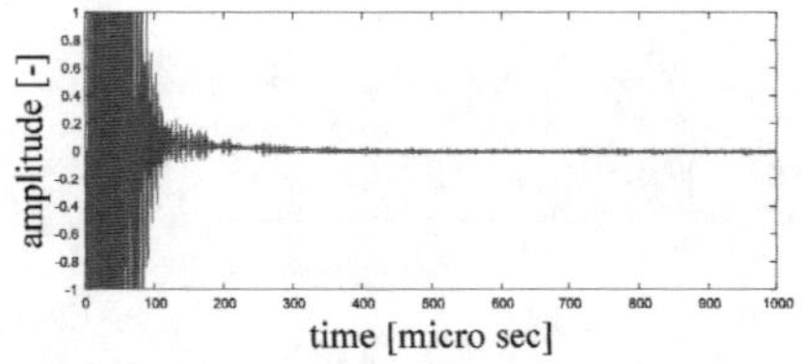

c. $H = 40$ mm

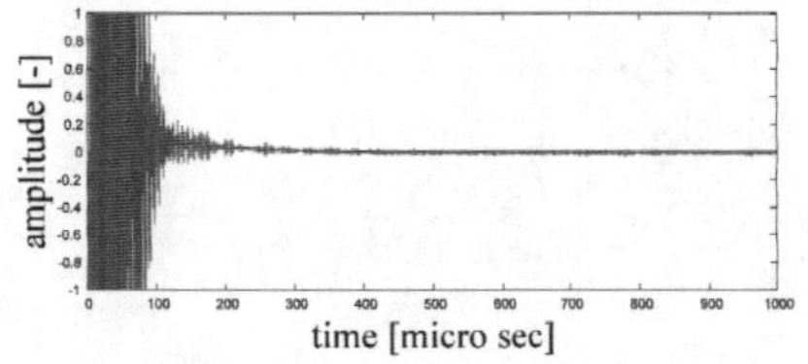

d. $H = 60$ mm

Fig. 11. Echo signal of the simulated fuel debris reflector measured by sensor A.

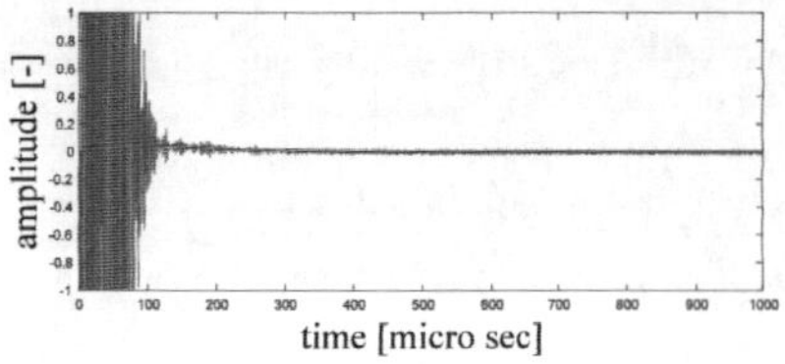

a. $H = 10$ mm

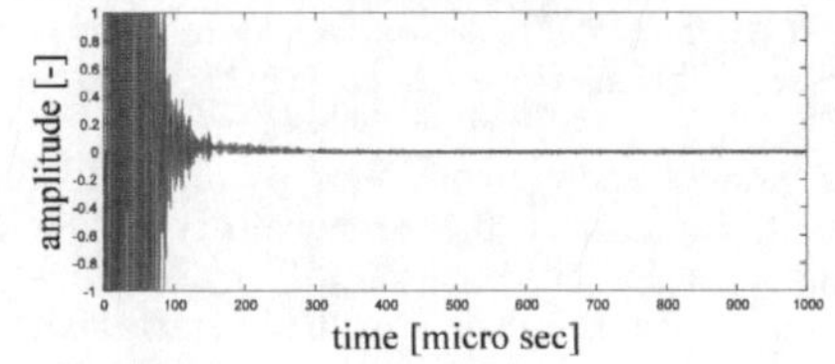

b. $H = 15$ mm

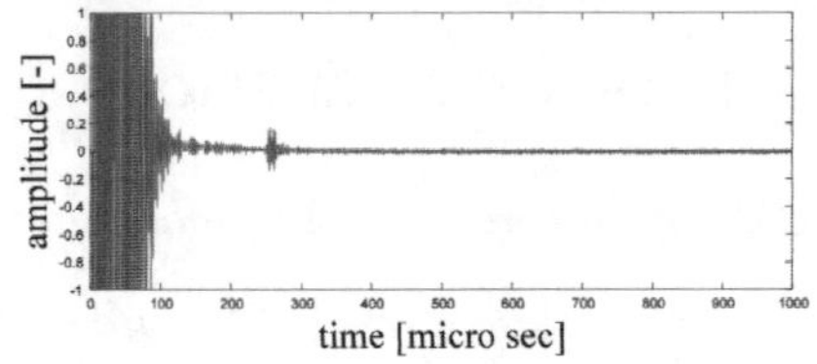

c. $H = 40$ mm

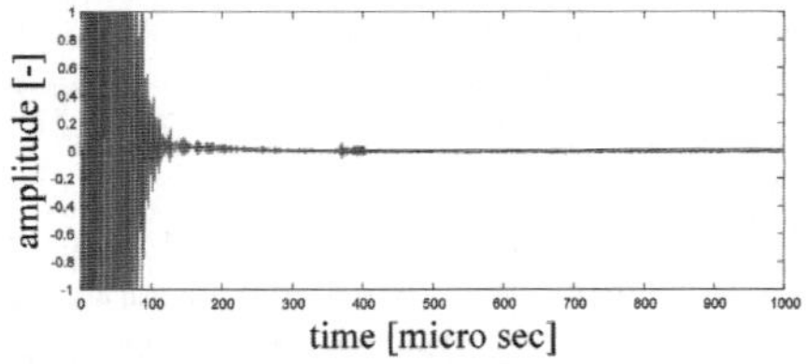

d. $H = 60$ mm

Fig. 12. Echo signal of the simulated fuel debris reflector measured B.

3.2 Image reconstruction

By using time of flight information, it is possible to reconstruct a three-dimensional image. Echo signals are being obtained while the sensor is scanning on the *x-y* plane. The distance between sensor and target (*H*) is set 30 mm. Then, shapes of both reference reflector and simulated fuel debris are constructed. The image reconstruction is developed using MATLAB®R2016a. Results are shown in Fig. 13 and 14. Figure 13 shows the image reconstruction of reference reflector by sensor A and Fig. 14 shows the image reconstruction of reference reflector by sensor B.

Figure 13a and 14a show the reconstruction image of reference reflector in the *x-y* plane, the horizontal axis is a distance in the *x*-direction and the vertical axis is a distance in the *y*-direction. Red color means that the distance between sensor and reflector is short (in this case is equal to 30 mm, *H*) and blue color means that the distance between sensor and reflector is far (in this case is equal to 80 mm). It can be confirmed that the radius of reference reflector is 50 mm. In addition, depth of reference reflector is known by subtracting the maximum depth (blue color) with the initial distance *H*. Thus, the height of reference reflector is confirmed 50 mm. Figure 13b and 14b show the bird's eye view of reference reflector using sensor A and B respectively. It can be seen that three-dimensional shape is fully obtained. Furthermore, image reconstruction results measured by sensor A and B show similarity shape.

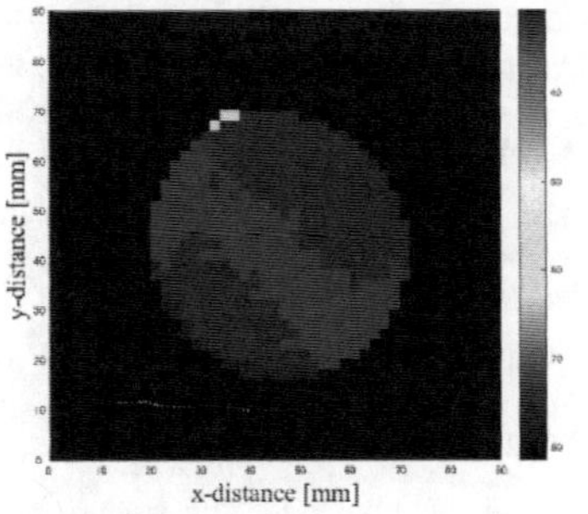

a. Image reconstruction in the x-y plane.

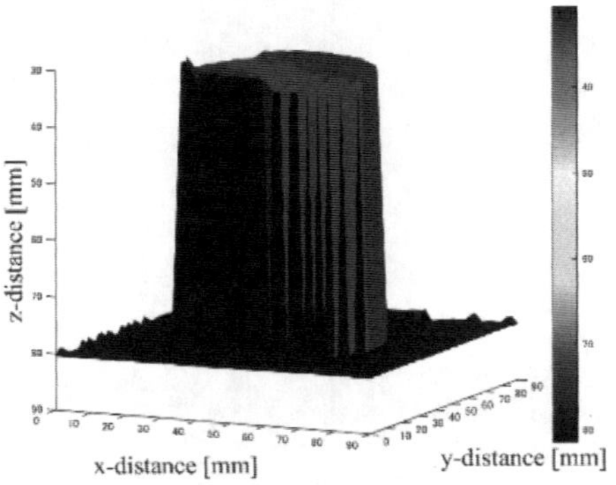

b. Image reconstruction in bird's eye view.

Fig. 13. Image reconstruction of reference reflector using sensor A.

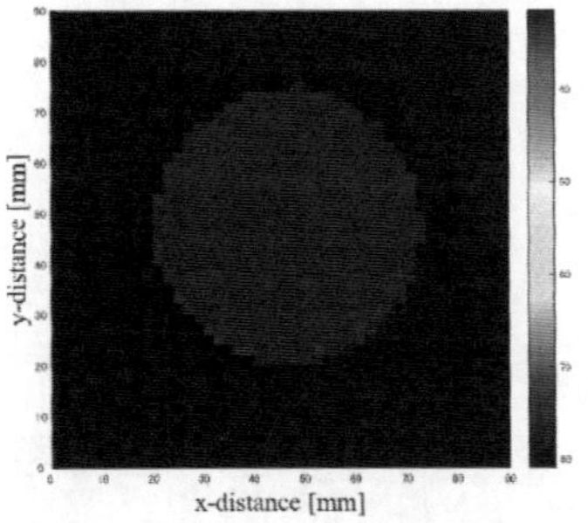

a. Image reconstruction in the x-y plane.

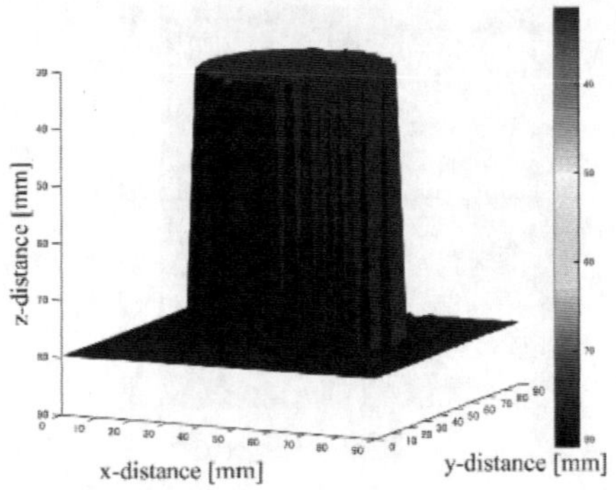

b. Image reconstruction in bird's eye view.

Fig. 14. Image reconstruction of reference reflector using sensor B.

Figure 15 shows the photograph of simulated fuel debris in the *x-y* plane (top view). This simulated fuel debris is scanned in *x-y* direction using both sensor A and B. Figure 16 shows reconstruction image of simulated fuel debris measured by sensor A in the *x-y* plane and Fig. 17 shows reconstruction image of simulated fuel debris measured by sensor B in the *x-y* plane. It can be seen that reconstruction image using sensor A is not clear compared with sensor B. It is because ultrasound's echo signals are greatly attenuated mainly due to scattering. As shown in preliminary experiment, it is difficult to get echo signals when sensor A is used in unstructured surfaced. Therefore, when sensor A is used, a distance between sensor and target surface can be calculated accurately. On the other hand, reconstruction image of fuel debris measured by sensor B has good accuracy. It can be seen in Fig. 17 that sensor B has a better result when target surface is unstructured.

Fig. 15. Simulated fuel debris in the x-y plane.

 https://doi.org/10.1051/e3sconf/20184201015

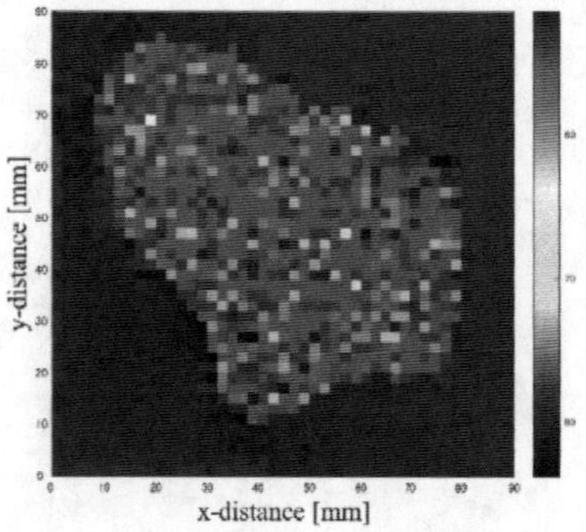

Fig. 16. Image reconstruction of simulated fuel debris sensor A in the x-y plane.

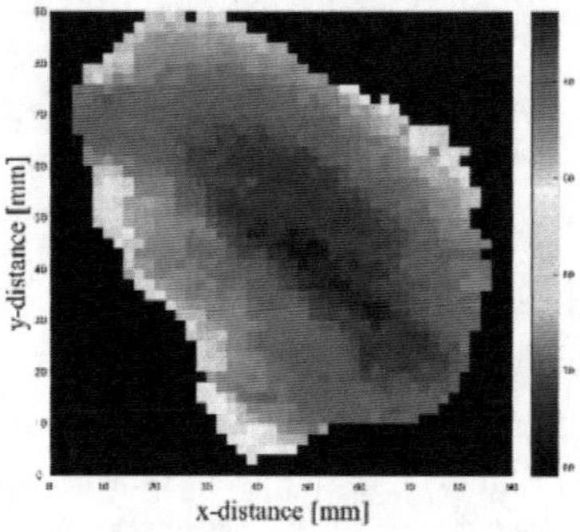

Fig. 17. Image reconstruction of simulated fuel debris using sensor B in the x-y plane.

4. Conclusion

Object visualization using air-coupled ultrasound is successfully developed. This measurement technique can be used as fuel debris inspection before removal the fuel debris is done. Hence, air-coupled measurement technique can be alternative technology in order to know the shape and size of fuel debris when the dry method is used. These are the following conclusions in this study.

1. The advantage of the air-coupled ultrasonic sensor was no need coupling fluid i.e. water. Thus, an acoustical mismatch between solids (transducers, test component) and air (coupling medium) must be taken care. This acoustical mismatch could be reduced by using matching layer.
2. Flat and focused air-coupled ultrasonic were good to get echo signals for flat target surface. However, the flat sensor had worse performance when the target surface was unstructured. It was due to ultrasound echo signals were greatly attenuated.
3. By using time-of-flight information, it was possible to reconstruct an image of the target surface. The spatial resolution of focus sensor shown good accuracy compared with the flat transducer. Therefore, it is recommended to used focused sensor when target surface is unstructured like simulated fuel debris.

References

1. A. Komori, *49th JAIF Annual Conference*, 1-33 (2016)

2. T. Shinkawa, Revised mid-and-long-term roadmap towards the decommisioning of TEPCO's Fukushima Daiichi NPP Units 1-4, Ministry of Economy, Trade and Industry (METI) workshop, pp. 1-23, (2013)

3. K. Takamori, *The 1st International Forum on the Decommissioning of the Fukushima Daiichi Nuclear Power Station*, 1-32 (2016)

4. H. Takegami, K. Takamatsu, C. Ito, R. Hino, K. Suzuki, H. Ohnuma, T. Okumura, Nippon Genshiryoku **13**, 7-16 (2014)

5. S. Takahashi, H. Ohigashi, Ultrasonics **49**, 495-498 (2009)

6. J.D. Fox, G.S. Kino, B.T. Khuri-Yakub, Appl. Phys. Lett. **47**, 465-467 (1985)

 https://doi.org/10.1051/e3sconf/20184201016

Development of Biobriquette from Nutmeg Seed Shells

Rukmana Idris[1+], *Ahmad* Tawiequrrohman Yuliansyah[1], *Suryo* Purwono[1]

[1]Department of Chemical Engineering, Faculty of Engineering, Universitas Gadjah Mada, Indonesia

Abstract. Nutmeg seed shell is one of biomass resources in Indonesia that has not been widely used. Nutmeg (Myristicafragans) in North Maluku province, especially in Tidore Island, is one of the largest incomes in the plantation sector. Nutmeg consists of pulp, mace, shell and seeds. Not all parts of nutmeg can be utilized or to be sold. The most widely used are the seed, mace (red salute that bind shell beans) and meat of nutmeg, while some parts have not been used optimally such as the shells. The abundance of nutmeg seed shells has only been discarded and not optimally utilized, therefore it needs to do a research on the utilization of nutmeg seed shells into briquettes that have good quality, which can be used as energy source and improve economic value. One fo the solutions is by converting it into bio briquette through pyrolysis process. Pyrolysis is chemical decomposition of organic materials through a process of heating without oxygen or under conditions of a little air, causing the release of volatiles and char formation. If the material contains water, at the beginning of heating, the free water will come out first at a temperature of about 120°C. Up to 170°C, volatile organic substances from hydrolysis will be released out. The conditions affect pyrolysis processes are temperature, heating time and particle size. The higher the temperature the higher the rate of decomposition.

1. Introduction

Bio-briquette are bio fuel substitute to coal and charcoal, mostly made of green waste and other organic materials, are commonly used for electricity generation, heat, and cooking fuel. Bio-briquettes are the solid fuel that compressed and have a specific shape. These compressed compounds contain various organic materials, including rice husk, bagasse, ground nut shells, municipal solid waste, agricultural waste. The composition of briquettes varies by area due to the availability of raw materials. The raw materials are gathered and compressed into briquette in order to burn longer and make transportation of the goods easier. These briquettes are very different from charcoal because they do not have large concentrations of carbonaceous substances and added materials. Compared to fossil fuels, briquettes produce low net total greenhouse gas emissions because the materials used are already a part of the carbon cycle.

Moisture content in the briquettes is between 10-20% by weight. Sizes of briquette vary from 20-100 grams. Some of briquettes are commonly known as pillows (oval), beehive (honey comb), cylinder, and others [1]. Some parameters affecting the quality of briquettes are calorific value, fixed carbon, moisture content, ash content and volatile matter. The higher the calorific value and fixed carbon, the lower the ash content, water content and volatile matter in briquettes is good quality briquettes. The parameters must fulfill of standards quality from SNI NO. 1/6235/2000 with moisture content $\leq 8\%$, ash content $\leq 8\%$, volatile matter $\leq 15\%$, fixed carbon $\geq 77\%$ and calorific value ≥ 5000 cal / g, [2].

Pyrolysis is chemical decomposition of organic material through a heating process with little or no oxygen or reagent. In this condition, the structure of raw material will undergo breakdown into the gas phase. Pyrolysis includes the case of thermolysis. Conditions affect of the pyrolysis process are temperature, heating time and particle size. Higher temperature will increase the rate of thermal decomposition.

Nutmeg seed shell is one of biomass resources in Indonesia that has not been widely used. Nutmeg (Myristicafragrans) in North Maluku province, especially in Tidore Island is one of the largest incomes in the plantation sector. The percentage wet weight of nutmeg parts is the meat of nutmeg (77.8%), mace (4%), shells (15.1%) and seeds (13.1%) [3].One of the factors of nutmeg can be utilized as a briquette is the contain of lignocelluloses which are hemicelluloses (9.51%), cellulose (28.63%) and lignin (50.14%).

During the pyrolysis process as result of heating, several reactions can occur. However, to simplify the

 https://doi.org/10.1051/e3sconf/20184201016

calculation, we use a modification of three-step mechanism [4].

Modification of three-step mechanism follows the equation:

$$\frac{dK}{dt} = -(k_1 + k_2 + k_3)\, K \quad (1)$$

$$\frac{dG}{dt} = k_1\, K \quad (2)$$

$$\frac{dC}{dt} = k_2 K \quad (3)$$

$$\frac{dPa}{dt} = k_3 K \quad (4)$$

Constant of reaction rate follows the Arrhenius equation as below:

$$k = A \exp\left(-\frac{E}{RT}\right) \quad (5)$$

2. Methodology

2.1 Preparation of Raw Materials

The raw materials are nutmeg seeds shells that dried under the sun. After that, the size of nutmeg seeds shells is diminished by a crusher and then sieved with a sieve size of 20 mesh.

2.2 Pyrolysis

Nutmeg powder was put into an oven at temperatures between 90 – 105°C, for about one hour to remove the remaining water. After that, the powder was put into a retort to be pyrolyzed at a temperature between 250 – 500°C. The pyrolysis time varied from 30 to 90 minutes. The products of pyrolysis were gas, liquid and solid. The solid was then removed from the reactor and mixed sticky with material and then pressurized up to 4 tons to form bio briquette. Fig. 1 shows the schematic diagram of the pyrolysis process.

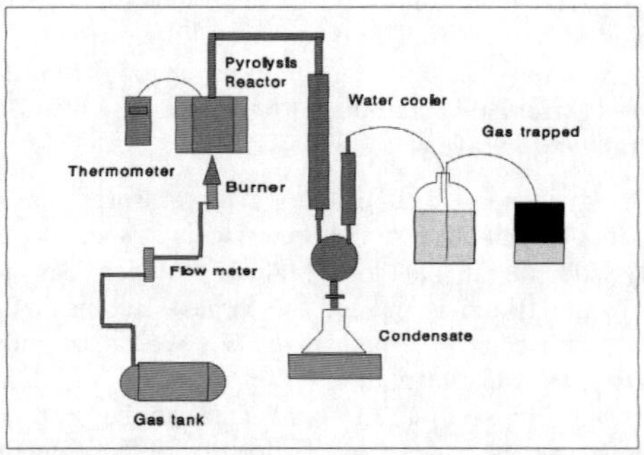

Fig. 1. Equipment for the pyrolysis experiment.

2.3 Briquetting Process

For the briquetting process, nutmeg seeds shells were pyrolyzed up to 350 ^{0}C, 400^{0}C, and 450^{0}C for 90 minutes. During pyrolysis process, the volumes of liquids and gasses were measured every 15 minutes. At the 60th minute of the pyrolysis process, gas samples were taken and stored in a vacuum tube to analyze its levels of CO, CO_2 and CH_4 contained. The gas pressure was measured using a manometer open. After pyrolysis, charcoal of nutmeg seeds shell was weighed 20 grams. Then it was mixed with adhesive with the variation % by weight of the adhesive with nutmeg seeds shells 10: 90%, 15: 85%, 20: 80%, 25: 75% and 30: 70%. After that, it was pressed into a cylindrical shape and then compressed by a hydraulic press with a compression pressure of 3 tons for 1 minute. Compressed briquette was then dried using an oven at a temperature of ± 60-80^{0}C.

2.4 Research Variable

Fixed variable consisted of pyrolysis time for 90 minutes and a particle size of 20 mesh. While the independent variable is the pyrolysis temperature of 350^{0}C, 400^{0}C and 450^{0}C and weight of adhesive mixtures of starch was 10%, 15%, 20%, 25%, 30%.

2.5 Analysis of the Results

Yield of products were analyzed using the following equation:

$$Y = \frac{W(G/C/Pa)}{Wo} \times 100\% \quad (6)$$

Analysis of the gas content of CO, CO_2, and CH_4 that comes out during pyrolysis was analyzed using gas chromatography (GC) on Instrumental Analytical Laboratory in Chemical Engineering Department UGM. In addition, to determine the quality of nutmeg seeds shells briquettes, several analysis were done such as calorific value, moisture content, volatile matter, fixed carbon, ash content at the laboratory of the Food and Nutrition in University Centre UGM. To determine the strength of the briquettes, we use shatter indexes test or drop test using ASTM D 440-86 R02 method.

3. Results and Discussion

3.1 Characteristics of Raw Materials

Briquette is a solid fuel that is made from organic material and has a specific shape. Based on the analysis results, nutmeg seed shells has an ash content of 1.119%, moisture content of 10.256% , volatile matter of 58.423%, fixed carbon of 30.712%, and calorific value of 4339.861 cal / g.

3.2 Effect of temperature to yield product

Thee effect of temperature to yield a product of gasses, liquids, and solids when the temperature of operation gets

https://doi.org/10.1051/e3sconf/20184201016

higher, the decomposition will increases. Yield product can be seen in Fig. 2.

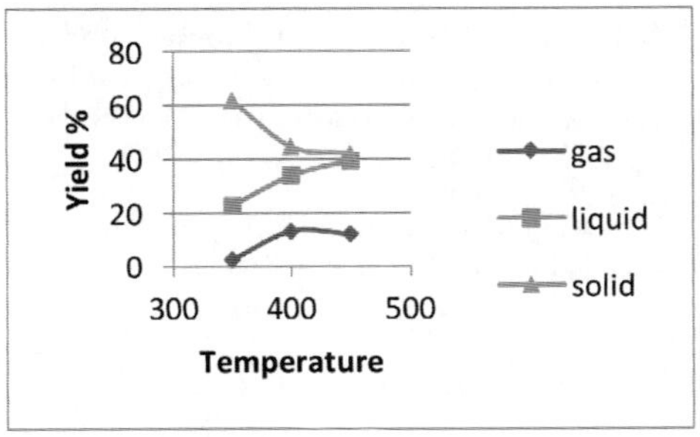

Fig. 2. Yield product

The rise in temperature will increase the rate of decomposition such as water content and volatile compounds, it causes fluid production increased at a higher temperature.

3.3 Pyrolysis Reaction Kinetics

The mechanism used is a three-step mechanism by Erawati et al (2013). This model depicts the solids mass reduction, the increase in mass of gas and liquid pyrolysis against time. Fig. 3 shows the comparison of gas and liquid by calculating the equation of reaction kinetics.

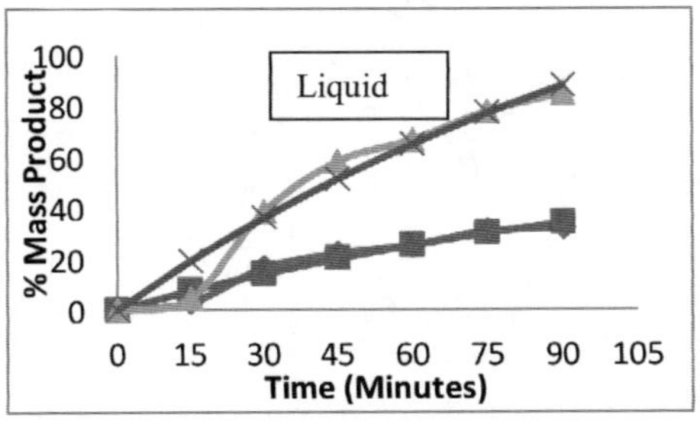

Fig. 3. Effect of time on the result of gas and liquid at temperature 350°C

According to Shankar et.al. (2010), the increased mass of gas and liquid is due to the breakdown of the chain on the lignocelluloses compounds that occur gradually. At the beginning of heating the short-chain will be degraded early and then the long chain would be degraded causing an increased mass of gas and liquids significantly.

The value of pre-exponential (A) and the activation energy (Ea) are k_1 (4.77 min-1, 27.89 KJ / mol), k_2 (0.518 min-1, 11.59 KJ / mol) and k_3 (6,36x10 -4 min-1, 11.22 KJ / mol). The values of activation energy of nutmeg seed shells briquettes are not much different from the activation energy of straw which is 11.3 (KJ / mol) [5].

3.4 Gas Analysis

Analysis of CO, CO_2 and CH_4 gasses in the pyrolysis process can be seen in Fig. 4:

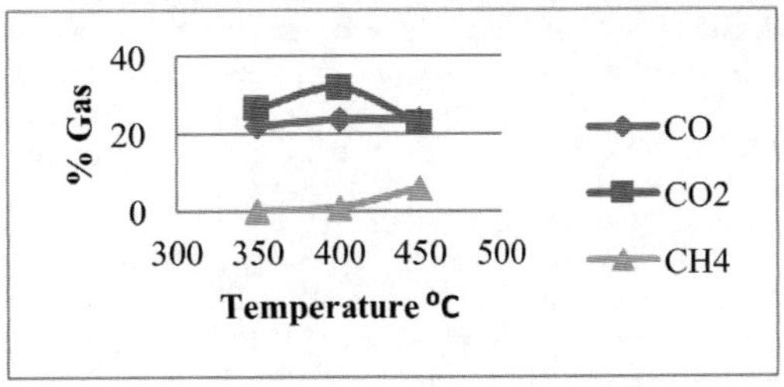

Fig. 4. Effect of temperature on gas analysis

According to Varfolomeev and Grachev (2015), the reason for increasing the production of CH_4 because of the lignin begins to decompose at above 350°C. The same thing for the production of CO, where it is increased when the temperature pyrolysis is high. These results are similar as reported by Ion (2013). However, the values of CO_2 is different, which increases at 400°C and decrease at temperature 450°C.

3.5 Quality of Briquette

The lower moisture content, ash content and volatile matter contained in the briquettes, the quality of briquettes increase. According to SNI No. 1/6235/2000 standard, the moisture content and ash content should be less than 8%, while the volatile matter is less than 15%. Based on the experiment the lowest value of moisture content before being mixed with the adhesive is at 400°C. This is different with the experiment at 450°C which has a value of water content is higher. In the study of Sastry et al. and Sengar et al. (2012) these things happen because it is influenced by the type of storage affect of the absorption of water from the air, and also causes the temperature of the air humidity which affects the water content. While the maximum ash content in 450°C is 11.252%, the value is not yet met the standard of SNI No. 1/6235/2000.

High value of volatile matter can degrade the quality of briquettes since it lowers the content of fixed carbon and can also affect the calorific value produced. Besides it will cause in increasing the amount of smoke produced from burning [6]. Low volatile matter contained in the results of 450°C is 8.385% without the addition of adhesives and already meet the ISO Standard No. 1/6235/2000.

Shatter index is a test to measure the strength of briquette due to impact. The results of the test show that shatter indexes at 400°C and 450°C of briquettes are less than at 350°C, it is also caused by high fixed carbon and the addition of adhesive as well. Higher adhesive content in the briquettes makes briquettes stronger, proves that the briquettes with 30% adhesive have the highest Shatter Index.

3.6 Environmental and Economic Aspects

Utilization of nutmeg seed shells as briquettes is one way to reduce the abundant of nutmeg seed shells that increase during harvest. Based on the literature, nutmeg production in 2013 is up to 225 tons and shells weighing 71.4 tons which are 28% of the production of nutmeg. The weight of the shells after converted into charcoal was reduced by 44.4%, is 31.7 tons of which were used as the raw material of briquettes. That shows that the utilization as briquettes can reduce waste of nutmeg seed shells.

While fro the economic aspect, people of Tidore mostly use kerosene and firewood as the main fuel for cooking needs. The results showed that the nutmeg seed shell briquettes have higher calorific value than wood with a relatively cheap price which is around Rp.4000 / kg.

4. Conclusion

1. The three-step mechanism depicts the solids mass reduction
2. When the temperature of operation is increased, the decomposition processes are getting better where the product of gasses and liquid increase.
3. The quality of the briquette partly fulfills the standard of SNI

Acknowledgements

The authors are grateful for funding from Directorate General of Higher Education (DP2M-Dikti) and from Chemical Engineering Department, Universitas Gadjah Mada.

Nomenclature

K	= initial of biomass
k	= Arrhenius
k_1	= Arrhenius constant for gas
k_2	= Arrhenius constant for liquid
k_3	= Arrhenius constant for solid
t	= time of reaction
A	= collision constant
E	= activation energy
R	= gas constant
T	= temperature
G	= gas
C	= liquid
Pa	= solid
W	= weight
Y	= yield

References

1. A.C. Brades, F.S. Tobing, Development of Briquette from Eichornia Crasipess Solm, Chemical Engineering Department, Sriwijaya University, Palembang (2007)
2. Paisal, M.S. Karyani, Quality Analysis of Bio-briquette from Durian Skin, Ambon Polytechnic (2014)
3. N. Nurdjannah, Nutmeg Processing Technology, Agriculture Department (2007)
4. E. Erawati. W.B. Sediawan, Modification of the Mechanism in Pyrolysis Kinetic Reaction of Bagasse, Chemical Engineering Department, Universitas Gadjah Mada (2013)
5. J.O. Adekunle, J.S. Ibrahim, E.I. Kucha, Br.J. Appl.Sci.Technol **7**, 114-123 (2015)
6. S.J. Veeresh, J. Narayana, Univ. J.Envi. Res.Tech **2**, 575-581 (2012)
7. M.K. Sastry, J. Bridge, A. Brown, R. Williams, Biomass Briquette : A Sustainable and Environment Friendly Energy Option for the Caribbean, Fifth International Symposium on Energy, Puerto Rico Energy Center-Laccei Puerto Rico (2013)
8. S.H. Sengar, A. Mohod, Y. Khandetod, S. Patil, A. D. Chendake, Intl. J. Energy Engineering **2**, 28-34 (2012)
9. J.S. Tumuluru, C.T. Wright, K.L. Kenney, J.R. Hess, A Technical Review on Biomass Processing: Densification, Prepocessing, Modeling, and Optimization, 3-5 (2010)
10. M. Varfolomeev, A. Grachev, Thermal Analysis and Calorimetric Study of The Combustion of Hydrolytic Wood Lignin and Products of Its Pyrolysis, Kazan National Research Technological University (2015)

Approach to Topographic Map, Hydrogeological and Environmental Isotopes Analysis to Delineation of Watersheds in Drought-prone Areas

Agus Budhie Wijatna [1] and *Tarsisius* Aris Sunantyo[2]

[1]Department of Nuclear Engineering and Engineering Physics, Faculty of Engineering, Universitas Gadjah Mada, Yogyakarta, Indonesia
[2]Department of Geodetic, Faculty of Engineering, Universitas Gadjah Mada, Yogyakarta, Indonesia.
[1]gusbudiftugm@gmail.com, [2]sunantyo@ugm.ac.id

Abstract. Gunung Kidul regency is one of the regencies in Indonesia that almost every year (often) is susceptible to drought, there is an even in term of synonymous with a drought in the province. Water is the essential lifeblood on Earth, with the power to generate, sustain, receive, and ultimately to unify life. Water is also one of the most important resources that need to be protected, conserved, and maintained to support humanity and all life on earth. Drought is a state of shortage of water supply in an area in which a prolonged period (several months to years). Usually, these events appear if a region is continuously subjected to rainfall below the average. Dry season long will cause drought. Drought is also a lack of water for human life and other living things in an area that is usually not a lack of water.

The environmental isotopes most widely used as a tracer in the study of hydrology cycle are the stable isotopes deuterium (2H), oxygen-18 (18O) and carbon-13 and the radioisotopes tritium (3H) and carbon-14 (14C). The use of environmental isotopes as a tracer in groundwater studies are usually complementary to approach topography, hydro geology, and geology which has been frequently used.

The result of this research show that elevation as an orthometric height of the 15 wells water in the research has a variation elevation which spread at most of the district in The from -18.390 m at Girisubo district up to 238.599 at Nglipar district m. Concerning with elevation of the water table at several geological cross section (East - West \) are close to orthometric height.**Keywords— Delineation; watersheds; drought-prone; tracer; environmental isotopes.**

1 Introduction

Gunung Kidul regency is one of the regencies in Indonesia that almost every year (often) is susceptible to drought, there is an even in term of synonymous with a drought in the province [1]. Water is the essential lifeblood on Earth, with the power to generate, sustain, receive, and ultimately to unify life. Water is also one of the most important resources that need to be protected, conserved, and maintained to support humanity and all life on earth. Drought is a state of shortage of water supply in an area in which a prolonged period (several months to years). Usually, these events appear if a region is continuously subjected to rainfall below the average. Dry season long will cause drought. Drought is also a lack of water for human life and other living things in an area that is usually not a lack of water. The regency is part of southern mountains tertiary Rocks [2]. The regency consists of 18 districts and geographical, is located at 110^0 21 'to 110^0 50' east longitude and 7^0 46' to 8^0 09' south latitude (see Figure 1). Based on geographic location and the existing ecosystem, the regency generally is an area that often experiences water shortages despite having relatively high rainfall.

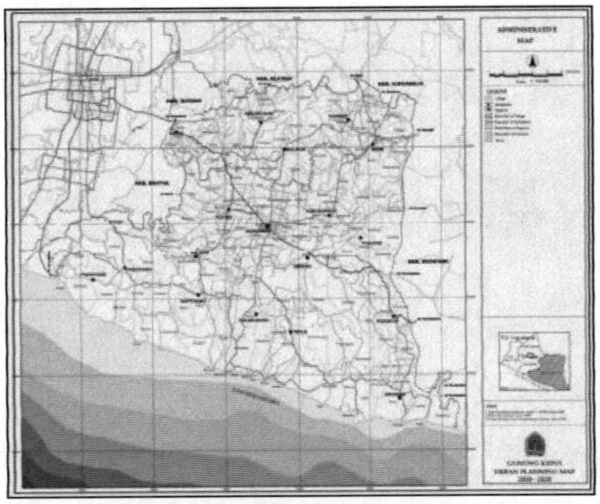

Fig. 1. Administrative maps belong to the regency [3]

All of the districts belong to the regency appears at Fig.1, it also appears that all of the roads which connect among districts. For rural communities, especially in karts mountainous areas, meeting water needs by relying on groundwater sources is highly unlikely. Concerning with clean water facilities and infrastructure barriers, the regency is mostly in a hilly area so it requires a big money to carry out investment for infrastructure to service water for community people in all of the district. The government regency has still tried to get too much money for servicing clean water for the community. In the regency, there are many underground rivers, but to take an advantage of the water required huge costs and high technology. Water sources in the regency until today has been exploited by water wells, springs, water from PDAM network, reservoirs or ponds and rainwater collected from a roof into storage tanks. Most sub-districts of the regency nearly years of drought, especially in the dry season. In the rainy season, all of the residents need water from filled springs, wells, water ponds and rainwater. Dug wells with a depth of over 10 meters widely used of the population in the district of the regency. Springs scattered in the district widely used to meet the various needs of life. In the dry season, dug wells, springs, and even some bin-bin rainwater become dry causing drought for citizens.

Fig. 2. Land use map belongs in the regency [3]

There is causes droughts such as the thin soil layer; groundwater; texture rough ground; climate; vegetation and topography. The low or high topography of a region affects the soil water content owned. Usually, the low-lying areas will have a soil water content which is more than in the highlands.

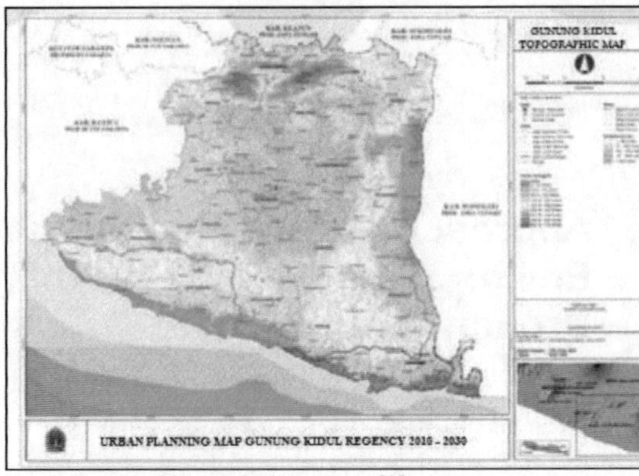

Fig. 3. Topographic map in the regency [3]

Due the water is absorbed by the soil which will flow from a high place to low place. Therefore because of the water will be more absorbed by the soil in the lower plains.

The Fig. 3 tells that green color is a forest. It means that coverage area of forest is lesser than other land use. Based on the topography, the regency can be characterized to be three zones above sea level. The zones are a north zone with has an orthometric height between 200 up to 700 m, the central zone between 150 up to 200 m and the south zone between 0 up to 300 m [4]. Concerning with Figure 3, a variation of colors at topographic map indicate the elevation of the area, it tells that red color is the highest elevation. Variation of colors at every district is different from place to place. The Variation of colors has changed gradually from red to green colors. It means that green color is the lowest elevation which is close beach as the long beach in the regency. [5] states that the topography determines whether there is an active movement of water through decay material. The role of the topography is as a motion controller groundwater upright (vertical), and it also affects the process of leaching and surface erosion that will determine the rate of removal of the result of weathering.

Delimitation and extensive water catchment areas with the approach of topography, geology, and hydro geology has been done by some researchers, however such an approach cannot be sure whether the water in the springs truly comes from catchment areas are being researched or derived from local rain come from the aquifer around or comes from catchments on it.

2 Approach/Methodology

There are several technologies to study watersheds in drought-prone. One of the technologies for the study delineation of watersheds in drought-prone areas is able to be done by using environmental isotopes tracer method. The research area as a case study is in several districts of the regency. In the science of hydrology, basic flow known as base flow or flow mainstay and role important as the only component of water supply (discharge) during the dry season.

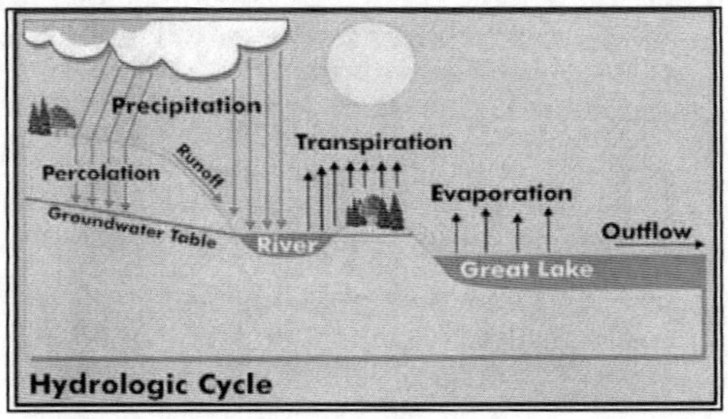

Fig. 4. Water continually recycled and returned to the ecosystem through the hydrology cycle [6]

Concerning with evapotranspiration in the regency as an average during in the year of 2007 is shown in Table1.

Table 1. Evapotranspiration in the regency [3]

Months	Rainfall (R) (mm)*	Potential Evapotranspiration (PE) (mm)	Actual Evapotranspiration (AE) (mm)	Water Surplus R – AE(mm)
January	272	140	106	166
February	339	128	106	233
March	229	137	112	117
April	232	130	105	127
May	383	121	83	300
June	30	110	65	-35
July	0	119	67	-67
August	0	132	69	-69
September	0	142	74	-74
October	215	152	79	136
November	510	142	79	431
December	238	142	94	144
Annual	2448	1595	1039	1409

Watershed hydrology (see Fig. 4 and Fig. 5) is a field of study that concerns itself with questions of where the water goes
when it rains, what flow paths the water takes to the stream, and how long water resides in the watershed.

From the Table 1, it tells that rainfall from January up to December is very different and has a big variation. Annual rainfall is 2448 mm per year. Potential evapotranspiration does not have a big different variation.

Water surplus from January up to December has a big variation between -74 mm up to 233 mm and its water surplus annually is 1409 mm. The average rainfall in the regency in the year 2007 of was 1720.86 mm/year with the number of rainy days on average 115 days per year. In the wet season of months are 4 up to 6 months, while the dry season of months is from 4 up 5 months. The rainy season begins in October up to November and ends in May-June every year. Rainfall peak season is in December up to February. The north regency is a region which has the highest rainfall over central and southern regions, while southern areas at the end rain. Temperatures of the regency for a daily average temperature is about 27.7 °C, where minimum temperature is about 23.2 °C and the maximum temperature is about 32.4 °C. Relative humidity in the regency has a range between 80% up to 85%. The relative humidity in the regency is not too affected by the high places, but it is influenced by season. The highest humidity occurs in the months January to March, while the lowest in September.

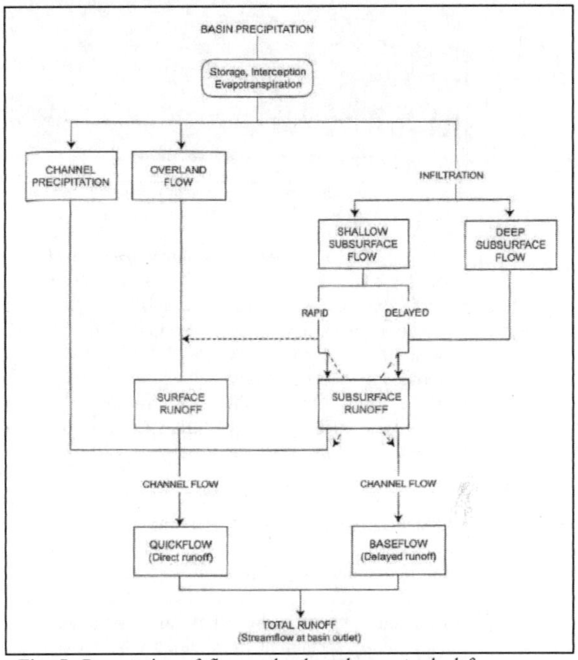

Fig. 5. Presentation of flow paths through a watershed from precipitation to stream flow at the basin outlet [7]

Methodology to carry out research consists of field survey and laboratory experiments. Field survey is aimed to look for the well pump which spread in the regency. Every location of the well pump has been taken data several data such as geodetic position, temperature, deep of good pump and additional explanation about utilities etc. Condition of groundwater in the regency can be seen in Fig.6.

The field survey was carried out about one month (three weeks in March and one week in April 2016) for 15 locations. Survey methods in the field by taking well water by pumping samples are 15 districts (see Figure 2). Names of the 15 district are Playen, Paliyan, Nglipar, Gedangsari, Tanjungsari, Nglipar (two points), Girisubo, Ponjong, Saptosari and Wonosari (four points). At the 15 selected locations as a sample of this research at the 15 districts, raw data were taken such as position (geodetic coordinate system), temperature, sample of water from the well pump and deep of the well pump.

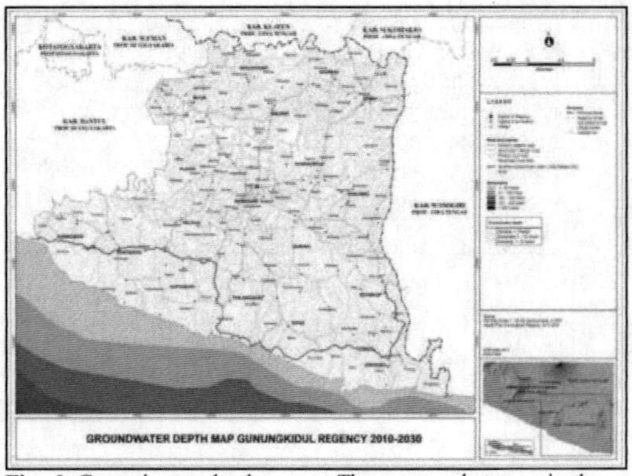

Fig. 6. Groundwater depth map at The regency between in the year of 2010 up to 2030 [3]

Laboratory experiments have been carried out in several steps. The first step of data processing is to analyze an assessment of watershed hypothesis of delineation using a topographical map. The second of data processing is to carry out to analyze assessment of watershed hypothesis of delineation using a cross section of the geological map of the research area. The last step of laboratory experiments is to analyze the isotopes ratio ($^2H/^1H$ or $^{18}O/^{16}O$) of the water sample which determined by mass spectroscopy at National Nuclear Energy Agency of Indonesia- BATAN Jakarta to get the boundaries of water catchment area. In the laboratory experiments, several analysis was done such as computation for transformation of the 15 points from geodetic coordinates system to be UTM coordinates system and orthometric height system, plotting of the 15 point at topographic map, Orthometric height (H) can be obtained by leveling and gravity observations and it has a physical meaning and depends on the gravity field of the earth. The orthometric height (H) is the length of the plumb line between the station P and P_0 at the geoid. Usually, orthometric height is measured by spirit leveling. H is also called the "height above Mean Sea Level".

2.1 Height System

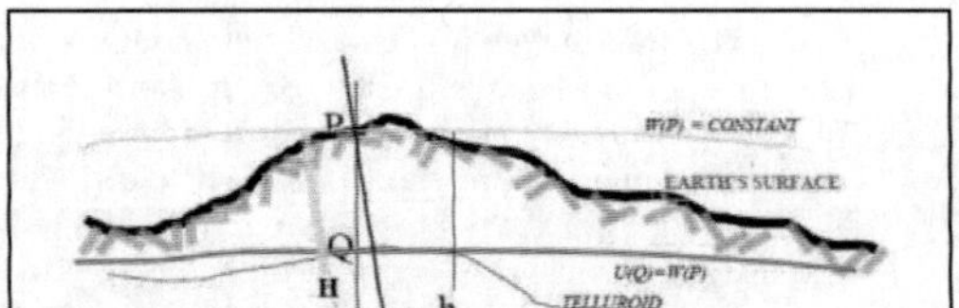

Fig. 7. Height system after modified [8]

Concerning with Fig. 7, it can be seen that geoid undulation (N) is the difference between ellipsoidal height (h) and orthometric height (H). The geoidal undulation (N) is defined as the height of geoid above (positive) or below (negative) the defined reference ellipsoid [8]. Ellipsoidal height (h) is the plumb line from the observation point at the topographical surface to the ellipsoid, and h is measured using GPS receivers. Ellipsoidal height (h) is a pure geometrical value. The local geoid is prepared by presenting the distribution of in contour form. This is called the (local) geometric geoid or gravimetric geoid. GPS will provide a direct solution if observed at leveling points with orthometric height information provided through connecting to tide gauges with a simple formula:

$$H = h - N \tag{1}$$

where h is the ellipsoidal height from GPS observation, H is orthometric height as computed from leveling network adjustment from tide gauges, and N is the geoidal height. Karts hydrology in the basin coalition river surface and subsurface river Bribin to Baron, the regency that output discharge volume of water availability can be known to be used in the management of water resources. The basin coalition consists of sub river basin district surface (non karts) and subsurface sub-watershed (karts). Debit of basin coalition as a continuous function of time is still unknown. The available data are very limited and temporary nature (event), but it is known that the output of perennial water. The basin coalition is a combination not yet well-defined system, given that most limits and extent cannot be ascertained as a result of a complex karts area. Sub-basin coalition aquifer is limestone surface has cracks and fissures, which are not as intensive as it is owned subsurface sub-watershed and not granular rocks. In subsurface sub-watershed not found the river's surface, but the surface of the minor groove formed in the valley between the hills during the rainy season and the water disappeared through cracks, crevices, caves, which have characteristic karts area [9].

2.2 The Basic Theory Of Environmental Isotopes Tracer Method [10]

Waters molecules are built of two elements: hydrogen (H) and Oxygen (O), with general formula, is H2O. Hydrogen has two stable isotopes: protium (1H) with an abundance of 99.985 %, deuterium (2H) with abundance up to approximately 0.015 % and a radioactive isotope of hydrogen-Tritium (3H). Oxygen also has three stable isotopes: 16O (common), 17O (very rare) and 18O (rare); whereupon just 18O is of practical relevance. So several types of water molecules are 1H216O-light, most common water molecule, and 1H218O and 2H1H16O-heavy, a rare water molecule. Due to chemical and physical processes, the isotope ratios can be changed (isotope effects). The, therefore, ratio of isotopes is not constant. The isotopic composition of water, determined by mass spectroscopy, is expressed in comparison to the isotopic composition of ocean water, called Standard Mean Ocean Water (SMOW). [11] states that the isotopic composition of water is expressed in per mil ‰ deviations from the SMOW standard. The deviation of deuterium is written 2H/1H and 18O/16O for oxygen-18. [12] The isotope composition of ocean water are well mixed and of rather a uniform composition. In the process of evaporation, an isotopic fraction occurs because the light water molecules are more efficiently evaporated than the heavy ones. Thus, water in the clouds isotopic-ally lighter than ocean water. In contrast, in the process of condensation from the cloud, during rain formation, a reversed fraction of isotope occurs because the heavier water molecules condense more easily than the light ones, leaving the cloud residual vapor depleted of 2H/1H and 18O/16O values. [13] As results, the vapor in the clouds is enriched in light water molecules than the ocean that supplied the water, with relatively negative 2H/1H and 18O/16O values. In contrast, the residual water phase becomes relatively depleted in light water molecules, more positive 2H/1H and 18O/16O values. [14] states that the basic theory of environmental isotopes tracer method essentially bases on these natural water molecules variations.

2.3 Delineation Of Watershed Using Environmental Isotopes

The first and most importance question when we want to conserve water catchment area are, where are the boundaries of water catchment area? First, hypothetical catchment area determined by synthesizing topographic, geological, hydro geological and hydro logical data. Then, hypothetical catchment areas were verified by environmental isotopes tracer method. Environmental isotopes tracer method is not to replace the delineation methods that have been used but is a complement to the methods.

3 Obtained / Expected Results

After field survey has been carried out and preliminary result concerning with laboratory experiments, finally several results has been achieved such as 1). Elevation (as a geometric height) of the 15 well pump. The 15 well pump are at Playen (geometric height : 237 m), Paliyan (geometric height 192m) , Nglipar (geometric height 265m and 218m), Gedangsari (geometric height 211m), Tanjungsari (geometric height 4m), two points at Girisubo district (geometric height 4m and 7m), Ponjong (geometric height 70m), Saptosari (geometric height 54m) and four points in Wonosari district (geometric height respectively are 231m, 200m, 192m, and 189m). Geodetic coordinates to be plotted on the land use map are shown in Fig.8 and Fig. 9.

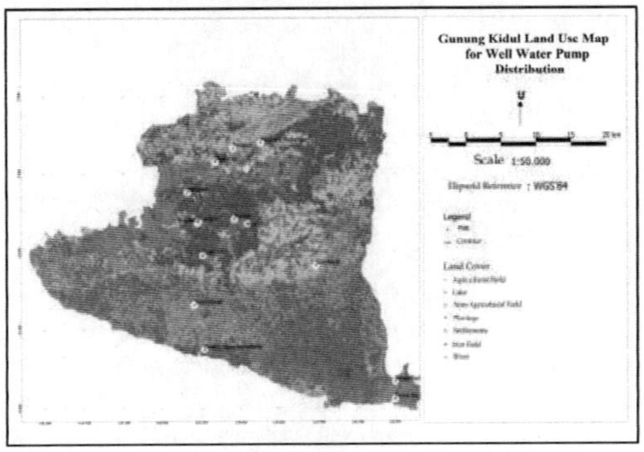

Fig. 7. Height system after modified [8]

2). Geodetic position. GNSS receiver was used to look for positioning such as geodetic coordinates (geodetic latitude, geodetic longitude, and geometric height). The receiver is also used to find positioning with UTM coordinates system for every well pump. 3). Explanation of the well pumps. Distribution using elevation for the 15 well water pump in the regency can be seen in Figure 8. Concerning with the Figure 7, plotting all of well water pump is spreading from north to south direction for the 15 districts. Concerning with orthometric height system, the result area consists of 15 points that have a variation elevation and spread at most of the district in The. 2). Deep water of the 15 well water by pumping are Playen (NA), Paliyan (NA), Nglipar (125m), Gedangsari (110m), Tanjungsari (80m), two points at Girisubo district (4m and 80m), Ponjong (67m), Saptosari (70m) and Wonosari (40m, 40m, 80m, and 60m). 4). Based on the function of the 15 well water pump that only 9 well water are in good conditions (it is still used to get water for people). 5). Watershed delineation using a cross-section of the geological map of the research is still in progress and 6). Watershed delineation using tracer hydroisotop over the research is still in progress.

Formula (1) must be used to process orthometric height. Two parameters of carrying out orthometric height are geoid undulation (EGM2008 as a subsurface map) and geometric height (it was obtained by GNSS receiver observation).

The geological cross section is very useful to identify groundwater. Concerning the result of field campaign for every deep well water, those can be used to plot geological cross section. The cross-section of geology describes the distribution of rock laterally and vertically along the geological cross section line. The cross-section of geology was made at any points observed wells and used in the analysis with a trajectory line that passes through the points of the well.

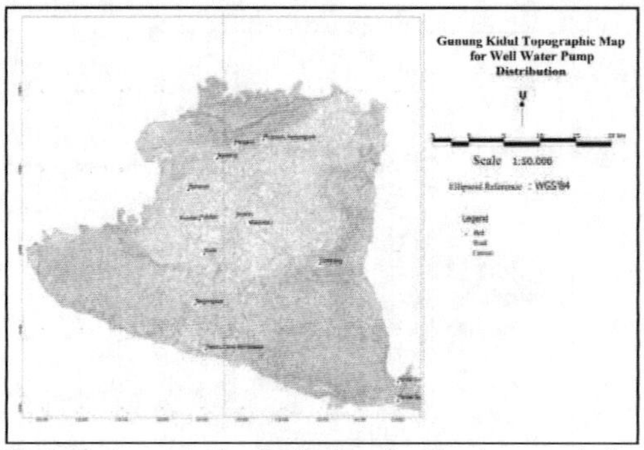

Fig.9. Contour map for distribution of well water pump in the regency.

Geological cross section was made by using a topographical map that is appropriate scale geological maps. Geological data based Regional Geological Map Sheet Surakarta - Giritontro with a scale of 1: 100,000 made by [15]. Geological data used from the form of data map layers of rock slope , rock types , geological structure and lithology of data distribution. The orthometric height of well pump and groundwater level obtained from field campaign then cross section geology is possible to be made together with groundwater levels sub surface conditions laterally and vertically.

Table 2. Computation Orthometric Height With Respect to the 15 Well Pump

Village	District	Latitude (D)	Longitude (D)	Orho. Height (m)
Banaran	Playen	-7.923805556	110.5330556	211.007
Sodo	Paliyan	-8.0045	110.5524444	166.268
Ngalang	Gedangsar i	-7.884972222	110.5686667	184.783
Pengkol	Nglipar	-7.868305556	110.5908611	191.682
Mojosari, Kedungpoh	Nglipar	-7.861694444	110.6265	238.599
Baron, Desa Kemadang	Tanjungsa ri	-8.125611111	110.5542778	-21.037
Tanjungsari	Tanjungsa ri	-8.068527778	110.5411667	-21.396
Beach of Sadeng	Girisubo	-8.189388889	110.8001944	-21.194
Beach of Sadeng	Girisubo	-8.167027778	110.8003889	-18.390
Gombang	Ponjong	-8.017888889	110.6970556	43.909
Ngerenehan	Saptosari	-8.238333333	110.5177778	30.054
Pulutan	Wonosari	-7.963611111	110.5452778	163.129
Pulutan	Wonosari	-7.961666667	110.5469444	166.118
Wonosari	Wonosari	-7.964722222	110.60925	173.995
Siyono	Wonosari	-7.9585	110.5925	205.009

The second step of this research is to produce geological cross section. At the beginning plotting of points after field, campaign using geodetic coordinates and deep well water.

Geological map of the regency after the 15 deep well water is shown in Fig. 10.

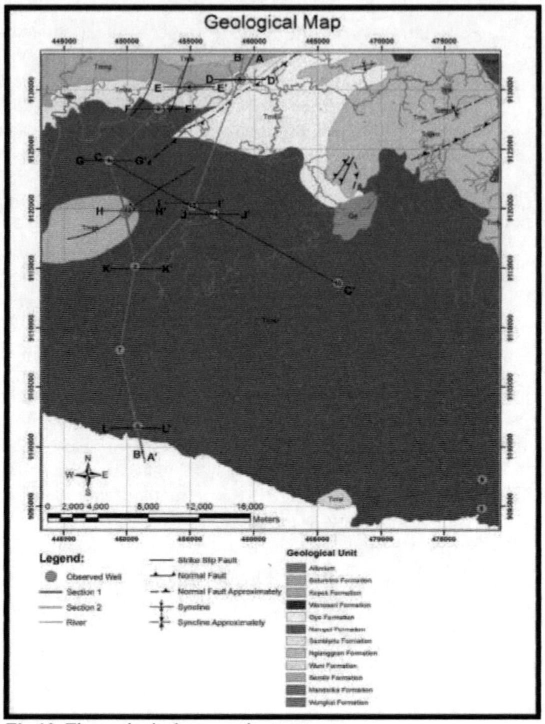

Fig.10. The geological map at the regency

To have a groundwater map, it is very important to carry out north to south geological cross section. The result of the geological cross section with the north to south direction is shown in Fig. 11.

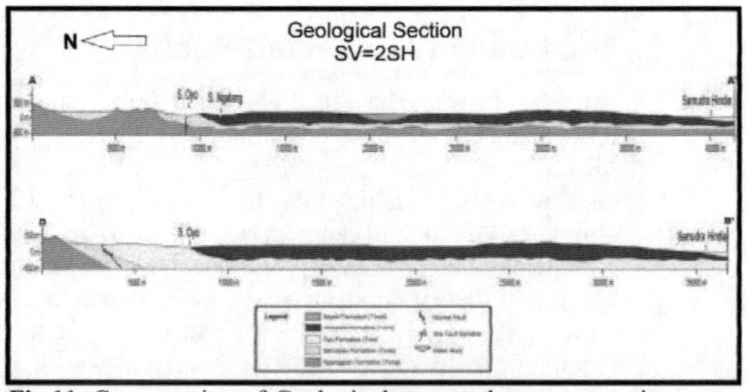

Fig.11. Cross section of Geological map at the researcg using north to south direction

After the geological cross section with north to south direction was done, then a geological cross section with west to east direction was carried out. The result of the geological cross section with west to east direction is shown in Fig.12.

 https://doi.org/10.1051/e3sconf/20184201017

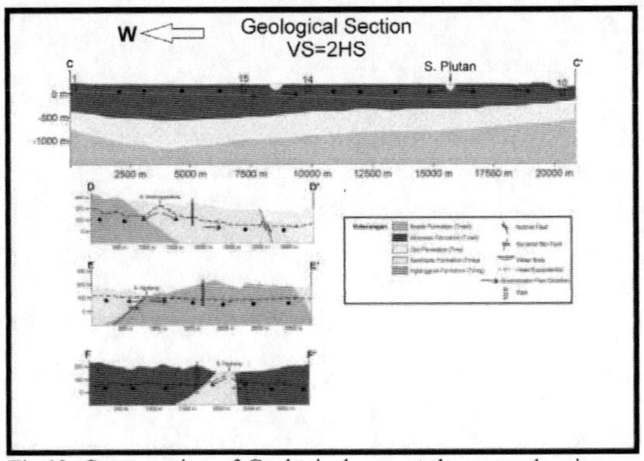

Fig.12. Cross section of Geological map at the research using west to east direction

Based on Fig. 10, a geological cross section of every point is very important to know about groundwater of this point, The geological cross section was made east-west direction and perpendicular with respect to the north-south direction of the regency. The result of every geological cross sections is visualized at Fig. 13.

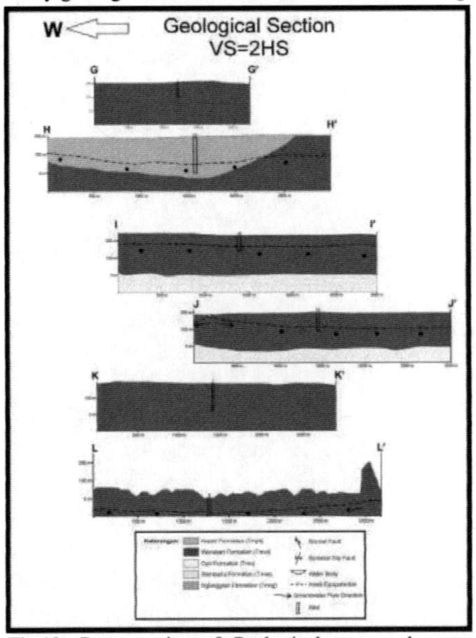

Fig.13. Cross section of Geological map at the research of every cross section perpendicular to the west-east direction

3 Conclusion

1. The 15 wells water in the research will be possible to be representative among the 18 districts for all of the districts.
2. Elevation as an orthometric height of the 15 wells water in this research has a variation elevation which spread at most of the district in The from -18.390 m at Girisubo district up to 238.599 at Nglipar district m
3. The deep water of the 15 well water pump has a variation from 4 m up to 125 m.Geological cross-sections (north-south direction and east-west direction) very use full to identify delineation of watersheds hypothesis in this regency.

References

[1] Rusmanto T., Mulyono, dan Irianto B.,2011. *Identifikasi gross β dan pengukuran parameter air di perairan Wonosari,* Prosiding Seminar Penelitian dan Pengelolaan Perangkat Nuklir, Yogyakarta.

[2] Rahardjo, W., 2007. *Foraminiferal Biostratigraphy of Southern Mountains Tertiary Rocks*, Yogyakarta Special Province. *Prosiding "Potensi geolog Pegunungan Selatan dalam Pengembangan Wilayah",* Yogyakarta 27-29 November 2007.

[3] Retnowati A., 2014. *Culture and Risk Based water and land Management in Karsts Area : An Understanding of Local Knowledge in The, Java,Indonesia*, Institute fur Geographie Giessen, Germany

[4] Husein, S. dan Sriyono. 2007. *Tinjauan Geomorfologi Pegunungan Selatan DIY/Jawa Tengah: Telaah Peran Faktor Endogenik dan Eksogenik dalam Proses Pembentukan Pegunungan.* Prosiding Workshop: Potensi geologi Pegunungan Selatan dalam pengembangan wilayah, Hotel Inna Garuda, Yogyakarta, 27- 29 November 2007.

[5] Grim, R.E. 1968. *Clay Mineralogy*. McGraw-Hill Book Company. New York, St. Louis, San Francisco, Toronto, London, Sydney, 596 p.

[6] Kresic, N. dan Stevanovic, Z., 2010. *"Groundwater Hydrology of Springs. Engineering, theory, management, and sustainability".* Elsevier Inc., USA. (3):91-104.

[7] Moritz H. and Wellenhof B.H., 2006.Physical geodesy, Springer Wien NewYork,

[8] Soenarto G., 2008. *Penaksiran Debit Daerah Pengaliran Gabungan Sungai Permukaan dan Bawah Permukaan Bribin - Baron Kabupaten The Propinsi DIY.* ITB., Abstract Dissertation, internet accessed on June 10, 2016.

[9] Kresic N and Stevanovic Z, 2010, Groundwater Hydrology of Springs, Butterworth-Heinemann, USA. pp 91-122 and pp. 312-316.

[10] Mazor Emanuel, 2011, Chemical and Isotopic Groundwater Hydrology, Second Edition, Marcel Dekker Inc., New York-USA; pp.168-194

[11] Richard W. and Healy, 2010, Estimating Groundwater Recharge, Cambridge University Press, 2010; pp.142-144.

[12] Mazor, Emanuel., 1997. *Chemical and Isotopic Groundwater Hydrology. The Applied Approach".* Mercel Dekker Ink. 2nd edition. USA. (9):168-194.

[13] MacDonald and Partners.1984. *Great Yogyakarta Groundwater study*, Directorate General of Water Resources Development of Republic of Indonesia.

[14] Surono, Sudarno and Toha B, 1992, Peta Geologi Lembar Surakarta-Giritontro, Jawa, 1:100.000, Puslitbang Geologi, Bandung.

AUTHOR INDEX